More advance praise for *What the Dog Knows*

"[Warren's] painstaking research on the history and science of working dogs debunks myths and explains what is known—and how much remains unknown—about canine abilities and behavior. By combining this hard information with anecdotes about training Solo, accounts of searching the North Carolina woods for dead bodies, and stories of other trainers and their dogs, she has produced a book that is both informative and entertaining. Although her love for Solo is palpable, she remains analytical and clear-headed, never romanticizing what he or other working dogs do."

—Bruce DeSilva, Edgar Award–winning
author of the Mulligan crime novels

"In a series of accounts that sometimes read like detective stories, Cat Warren . . . takes us through the steps needed to create dogs that search for people—both living and dead—while describing her life and her special bond with a German shepherd named Solo."

—Stanley Coren,
author of *Born to Bark* and *Do Dogs Dream?*

"*What the Dog Knows* is first the story of the relationship between a hard-working cadaver dog and his human companion. But that deeply felt relationship opens the way to an exploration of the working-dog world and, in doing so, becomes something more—a realization of the intelligence, determination, and decency of these animals, a story both wonderful and wise."

—Deborah Blum, author of *Love at Goon Park:
Harry Harlow and the Science of Affection*

"The capabilities of these specially trained working dogs are remarkable. The author provides fascinating insider information about a meaningful partnership that has important legal and personal consequences."

—Amy Hempel

"Cat Warren has captured both the magic and the best science behind the success of the modern working dog. This book masterfully shows how even the best technology cannot compete with our best friends. If you have ever wondered what dogs are truly capable of, this is the book for you."

—Brian Hare, evolutionary anthropologist,
director of Duke University's Canine Cognition Center,
and co-author of *The Genius of Dogs*

"It doesn't take a dog lover (such as myself) to appreciate Cat Warren's remarkable *What the Dog Knows*. Prepare to be enthralled and enlightened by this story of Solo and his mistress, whose clear, lively, personal, and intelligent writing will nail you from page one. It's a toss-up as to who is more fascinating—the dog people or the dogs themselves—in this wonderful and altogether unique book."

—Lee Smith, author of *The Last Girls*

"Working dogs, be they search-and-rescue, cadaver, or explosive detection specialists, are—like their human partners—a breed apart. They inhabit a world of complete commitment, utter dedication, and extraordinarily rigorous training. *What the Dog Knows* is greatly enriched by author Cat Warren's own love of digging. She and Solo take us on some fascinating detours through history and phony-baloney claims en route to the science, wonder, and awe that all rightly surround dogs' noses."

—Sue Russell, author of *Lethal Intent*
and *The Illustrated Courtroom*

WHAT THE DOG KNOWS

THE
SCIENCE
AND
WONDER
OF
WORKING
DOGS

CAT WARREN

A Touchstone Book
Published by Simon & Schuster
New York London Toronto Sydney New Delhi

Touchstone
A Division of Simon & Schuster, Inc.
1230 Avenue of the Americas
New York, NY 10020

First Touchstone hardcover edition October 2013

For information about special discounts for bulk purchases, please contact Simon
& Schuster Special Sales at 1-866-506-1949 or business@simonandschuster.com.

The Simon & Schuster Speakers Bureau can bring authors to your live event.
For more information or to book an event, contact the Simon & Schuster Speakers
Bureau at 1-866-248-3049 or visit our website at www.simonspeakers.com.

Designed by Aline C. Pace

Manufactured in the United States of America

10   9   8   7   6   5   4   3   2   1

Library of Congress Cataloging-in-Publication Data

Warren, Cat.
    What the dog knows : the science and wonder of working dogs / Cat Warren.
        pages   cm
    "A Touchstone book."
1. Search dogs—Anecdotes.   2. Warren, Cat.   I. Title.
    SF428.73W37   2013
    636.7'0886—dc23   2013012006

ISBN 978-1-4516-6731-8
ISBN 978-1-4516-6733-2 (ebook)

To David, my one and only

Contents

CONTENTS

WHAT THE
DOG KNOWS

I crossed a moor, with a name of its own
And a certain use in the world no doubt,
Yet a hand's-breadth of it shines alone
'Mid the blank miles round about:

For there I picked up on the heather
And there I put inside my breast
A moulted feather, an eagle-feather—
Well, I forget the rest.

—Robert Browning, "Memorabilia," 1855

Introduction

I've grown more comfortable working with the dead. With parts of them, really. A few teeth, a vertebra, a piece of carpet that lay underneath a body. One of my German shepherd's standard training materials is dirt harvested from sites where decomposing bodies rested. Crack open a Mason jar filled with that dirt, and all I smell is North Carolina woods—musky darkness with a hint of mildewed alder leaves. Solo smells the departed.

Solo is a cadaver dog. I occasionally get a call asking for our services when someone is missing and most likely dead. People have asked me if Solo gets depressed when he finds someone dead. No. Solo's work—and his fun—begins with someone's ending. Nothing makes him happier than a romp in a swamp looking for someone who has been missing for a while. For him, human death is a big game. To win, all he has to do is smell it, get as close as he can to it, tell me about it, and then get his reward: playing tug-of-war with a rope toy.

I never thought death could have an upside. I certainly never expected a dog to point that out to me. Since I started training and working with Solo eight years ago, he's opened a new world to me. Sure, some of it is dark, but gradations of light filter through so much of it that I find it illuminates other spaces in my life.

Solo and I have different reasons for doing this work. What appears to motivate him is not just the tug-toy reward at the end (although that pleases him greatly) but also the work itself, as he sweeps a field like a hyperactive Zamboni on ice, tracking will o' the wisps of scent down to their source. What motivates me is watching Solo, a black-and-red shepherd with a big grin and a huge rudder of a tail. He captures the hidden world his nose knows and translates that arcane knowledge for us humans. As one of the K9 unit sergeants said, admiring Solo's clear body language, "You can read that dog like a book." An easy book, happily, for a working-dog beginner like me. More Dr. Seuss's *One Fish Two Fish* than James Joyce's *Finnegans Wake*. It's a good thing that Solo's approach is Seuss-like, because the larger landscape of the missing and dead sometimes keeps me up at night pondering, poking at small details, trying to understand an unknowable plot. As one famous cadaver-dog trainer said, "Search is the classic mystery."

My hobby can raise eyebrows. While close friends and a few of my university colleagues embraced the idea with delight, others cringed. With some colleagues, I knew better than to mention it. Mostly, they don't know, as there's no reason to. One administrator, surprised when I told him I had to miss an upcoming faculty meeting to take Solo on a last-minute homicide search, came back to me the next day. Perhaps, he suggested with laudable optimism, I could put cadaver-dog work on my curriculum vitae as extension and outreach? I am not sure this peculiar avocation burnishes my academic credentials. I appreciated his willingness to consider it, though. I know cadaver dogs are an esoteric branch off the working-dog tree, as well as an acquired taste. If someone turns up her nose, I change the subject to politics.

Academics, of course, don't have a monopoly on passing judgment. During a moment of calm at searches, sometimes a sheriff deputy or police officer will ask about what I do for a living. When I tell them

I teach at a university, some wince as well, eyeing me for signs of effeteness—and weakness. Then, temporarily at least, we forget about our differences and continue the search, where we are on common ground.

Solo has no idea that I have a split life, or that he's partly the cause of it. Why should he? He's a dog. He's unaware that human death and decay cause disgust or ambivalence. For him, death is a tug toy. For me, Solo is the ideal intermediary between me and death. When we search—but even when we train—he becomes the center of my universe, narrowing my scope to the area we're searching. My job is to guide him when needed but let him do his job independent of me, to make sure he has plenty of water and isn't too close to traffic or a backyard Rottweiler, and to watch him closely the entire time, as he tests the air currents and reacts to them.

Looking for a body is an idiosyncratic way of walking in the woods. If I come across a snapping turtle or see an indigo bunting flash in the trees, or if the winter woods open onto an abandoned tobacco barn surrounded with golden beech trees, the pleasure remains, though the reason for being there is a somber one. And it's not all beauty out there: The hidden barbed-wire fences, the catbrier and poison ivy, the deadfall, clear cuts, and garbage dumps that litter the woods all demand my attention, and they get it. Though Solo doesn't love pushing through briar, other than that, even in junkyards or abandoned homesteads, he enjoys sticking his nose into the dark hollows and spaces created by piles of rusted-out heaps and old foundations. I worry more about copperheads, jagged metal, and broken glass than I do about the dangers posed by people, even when a case involves homicide. I do know more about the drug trade in North Carolina than I did before, and I avoid certain truck stops along the I-40 corridor, even if the fuel gauge is near empty.

Overall, the world seems less frightening with a large dog at your side—and that is perhaps especially true when one faces death. For thousands of years, and in numerous religions, from Hinduism in

India to the Mayan religions in Mesoamerica, the dead have depended on the continued assistance of canines to help guide them wherever they are going. The Zoroastrians wanted a dog present at funerals, though not just any dog. Preferably a "four-eyed" dog, with a spot of darker fur above each eye. I imagine an ancient shepherd version of Solo doing a gleeful slalom through the mourners.

Tragedy, occasional incompetence, and inevitable cruelty are part of the work, a given. I don't forget those facets: They are relevant, but they don't shine, and not just because Solo is present. Savvy police and sheriff investigators, experienced search managers, locals who know every dirt road and creek in the county, and families and communities that care—because most do—end up occupying much of my selective memory space.

Working with this one ebullient German shepherd and his good nose was the beginning of an odyssey that has started to merge worlds I've loved separately for decades: nature, researching and writing about biology and applied science, and working and playing with animals— especially dogs. The dog's nose has led me to environmental biologists, forensic anthropologists, cognitive psychologists, medical examiners, and military researchers. I've been able to interview, meet, and apprentice with talented working-dog trainers and handlers—people I've ended up liking as much as I like dogs. I've trained alongside canine handlers and trainers who work with drug, bomb, and patrol dogs. In that world of law enforcement, dogs are not just good friends but irreplaceable extensions, lending noses and ears and sometimes bodies and teeth to their human partners, smelling and hearing things their human handlers cannot, going places most people are reluctant to go.

My epiphany was not that working dogs are miraculous—by themselves, they aren't—but instead, how inextricably linked their success is to the quality of their handlers, and the trainers who train the handlers. Working dogs' success is far from a given: It takes imagination, deep knowledge, and constant work to train and handle dogs who work with their noses for a living. These are the dog people whose lives and careers are so interwoven with working canines that it can be difficult to see where the person ends and the dog begins; they complete each other. Not because the work they do is smooth or easy. The opposite is true. Often they are working in dangerous environments, or in the midst of devastation—whether from crime, war, climate change, earthquakes, or airplane crashes. The rare perfection of that human and canine partnership in our weird, complex, mechanized world is what keeps working dogs from obsolescence. Working dogs are a holdover from simpler times. Sometimes they're seen as a sentimental and unnecessary indulgence. Not all dog-and-handler teams are effective. But when they are good, they are very, very good: They can distinguish scent, cover territory, and accomplish tasks that no machine is capable of. We have new needs for the old work of dogs.

I don't handle and train dogs full-time. I probably will always be a serious hobbyist. Despite the nightmares I have when I make errors, I still return. I'm hooked. As I get better at juggling university demands and training demands, and as I learn to deal with the inevitable sadness, what remains is the intense physical and mental challenge of stripping a search to its essential elements so the dog can do his best work. Walking in the woods with Solo, as scent starts to loft in the morning warmth, I can concentrate so fiercely on our surroundings that time slows and warps. Or I can simply enjoy a night of training as the fireflies come out and Solo waltzes through solving a complex scent problem, a dancing figure in the dark.

He is a dog who both lives and narrates as his brown eyes snap with pleasure and impatience and he comes bounding across a cow pasture to lead me back to what he has discovered two hundred feet away.

Hey, come here, will you? Quick. The dead stuff is over here. Let me show you.

The Little Prince of Darkness

Being an only child is a disease in itself.
—G. Stanley Hall, *Of Peculiar and Exceptional Children,* 1896

The German shepherd pup had to be lifted out of the slit in his anesthetized mother's womb. A heavy lump. A litter of one.

He had a gorgeous head and great strength for a newborn, said Joan, the breeder in Ohio, in her e-mail to me. His strength wasn't a huge surprise, as his mother's nutrition flowed to him with no competition. I gazed at the first post-C-section photos of him: nestled comfortably and solidly in Joan's cupped hands in one shot, latched on to one of his young mother's eight teats in another shot. He had his choice of milk dispensers. He looked squashy and squint-eyed. His head looked like a mole's, not gorgeous at all, although Joan would know better. This single pup was her twenty-fifth litter of shepherds. He would be Vita's one and only.

The pup had something in addition to looks, strength, and remarkable sangfroid for a newborn. He also had a fine nose, Joan wrote. "Yesterday, even within hours of coming home, he woke up when I entered the room and his nose was working scent!" I barely registered the irrelevant news. I knew abstractly what "working scent" meant, but it didn't interest me. I'd taught my previous two German shepherds to keep their big noses away from visitors' crotches. "No sniff" was a standard command in our house.

The most important news, the lead, was buried a few paragraphs into Joan's e-mail: "You have the choice to have our little prince as we see how he develops." She assured me that we could discuss any concerns I might have about his being a singleton, and that she—and her pack of adult shepherds—could help the pup overcome the issues he might have.

Concerns? Issues? David and I had just won the puppy lottery with a handsome, healthy male. We had a pup. I had been stalking my e-mail in-box for the last week, waiting for the birth announcement. It had been almost a year since our beloved gentle shepherd, Zev, had died. The next chapter of our life with shepherds had finally arrived. I ran to find David, working on his logic courses in the study. I flitted around the living room. I landed in front of the computer to read the entire e-mail aloud to him. David patiently stood and listened as I made the words real. I waited for my euphoria to dissipate before I e-mailed Joan back, so my tone would be mature and balanced. All that planning and work and cost and emotional investment for one lone pup instead of a squiggling mass of them. Others on the waiting list would be so disappointed with the news. I knew all of that. Then I gave in to being overjoyed.

I had fallen in love with this Ohio breeder's line of shepherds, and the idea of this pup, ten months before. Joan Andreasen-Webb bred and raised German shepherds from West German lines, nourishing her pups with goat milk, a raw-meat diet, and lots of early exposure to the world. Her adult dogs lay on the sidewalk under café tables; they attended children's reading hour at the library; they herded sheep and starred in a ring sport called Schutzhund that I knew little about, except that it involved biting on command. A couple of her pups even became police K9s. As a reporter decades before, I had done a ride-along with a police K9 and been both impressed and horrified by the dog's intensity and deep-throated bark. I didn't want that in a German shepherd. This pup was destined for two jobs: to lie quietly beneath my desk while I worked, then leap up and reign supreme in the obedience ring, a hobby I'd abandoned when Zev became too sick to compete.

I finally stopped daydreaming and looked up "singleton" on the web. In mathematics, a singleton is a set with exactly one element. In humans, it's the way most of us arrive, as a single newborn. In dogs, "singleton" means exactly the same thing, only with horror stories attached. The web is like that, though. You can look up the common cold, and the symptoms read like it's the plague.

Pups in standard litters give and receive thousands of signals from each other daily, as they tumble over one another, licking and biting, squealing in pain, pissing and licking in apology, and then easing up on the bite. The scrum of a litter gets a pup ready for the rough-and-tumble of the dog park, the next-door neighbor's snappy Chihuahua, and the chance encounters with weird people—and children. A singleton pup, though, lives in a universe of "yes." They tend to lack "bite inhibition." They have "touch sensitivity." They are "unable to get out of trouble calmly and graciously." (Although I wasn't an only child, I related to that last one.) They have an "inability to handle frustration." (That one, too.) Joan had told me about the potential upside, and I went on to read those sections with great relief. Singleton dogs can make extraordinary companions, as they bond closely to people. Sometimes.

David and I avoided the what-ifs that night. We had named this pup even before Vita came into heat: Coda, literally "tail" in Italian, the musical movement at the end of a composition—a looking back, a thoughtful reflection, a summation. This pup was going to complement our academic and social lives, not disrupt them. I recently had been granted tenure at a good university and was finally building up a head of steam to chug through academic life like the little engine that could, producing research and fulfilling my destiny as a spunky and hip faculty administrator who wore cool black outfits and could speak truth to power without compromising my principles. Nothing would stop my momentum. Perhaps I wasn't an academic superstar, but I was pretty darned good at what I did. A pup was a simple gift, my reward for that work, and a welcome distraction from what felt like an increasingly long university engagement.

We were realistic, or at least that's what we told ourselves. We expected this pup, from West German breeding lines, to be higher-energy and tougher than Zev, who mostly loved to lie in the grass and smell the flowers. We already had a dog who took some of our time and energy: a beautiful female Irish setter we had adopted from my father several years before to help him adjust to a new life with a lovely woman who wasn't accustomed to large and quasi-uncontrollable dogs. We offered to take Megan to lighten the dog load. I lied to David and told him it would be fun, a real adventure, not just a filial duty, to adopt a year-old Irish setter in heat.

Though Megan was now four and had graduated beyond those moments when we fantasized about placing her on a nice farm in the country, my feelings about Irish setters hadn't changed much since childhood. They had filled our small house in Oregon with their gaiety, their indifference to obedience, and their uncanny ability to bolt. They would disappear into the dark fog of the Willamette Valley, cross-country journeys to nowhere, ending up lost, miles away from our little house on the hill. Always at night. Their other sins were insubstantial: jumping on guests, snagging empty rolls of toilet paper to play with, occasionally slinking up onto beds and easy chairs when no one was paying attention. My father loved their minor mutinies, loved to stroke their silky setter heads. They distracted him from a grinding schedule: a demanding research career; a wife who, because she was paralyzed, needed nursing; and three occasionally wild children who needed raising. The setters and their escapades were his only vacation.

Unlike my father, I didn't want dogs as a distraction; I wanted dogs who would engage completely with me and vice versa. By my early twenties, I had settled on German shepherds as my favorite breed. Partly because I loved their intelligence and dignity and their physical resemblance to wolves, partly because they were the antithesis of setters. David met me when my second shepherd was still a young dog and fell in love with him. Zev was an easygoing ambassador for the breed.

David and I realized the squashy mole needed a name that suited him better than Coda. His entry into the world felt less like a tail end and more like something improvisational. So David, a lover of jazz, renamed him Solo.

Animal behaviorist and author Patricia McConnell, who has devoted a good portion of her career and research to dogs with behavior problems, has a chapter in one of her training books on anger management in dogs. She wrote about her reaction when her favorite border collie gave birth to just one pup: "I'm supposed to help people, not cause the very problems I'm trained to alleviate, so when the vet confirmed that the litter contained a total of one puppy I was beside myself. You might think that it wouldn't be a crisis but it felt like one to me." McConnell briefly considered euthanizing the pup before rejecting that idea as she held the small warm bundle of fur. "Over the years I have seen what appeared to be a disproportionately large number of singleton puppies with serious behavior problems." She was the dog behaviorist who knew too much. Nonetheless, she decided she would experiment. For the good of her research and perhaps the good of future clients who came to her in desperation over their singleton dogs.

When he was only five weeks old, McConnell wrote, the border collie pup growled at her in fierce aggression, lips curled back from tiny milk teeth. "All I had done was touch him."

. . .

You like me because I'm a scoundrel. There aren't enough scoundrels in your life.

—Han Solo, *The Empire Strikes Back*, 1980

Joan nicknamed the singleton pup "HRH," for His Royal Highness. Solo was the king of everything. He had the canine equivalent of an Exeter education before he was eight weeks old. Being a litter of one had its perks. Joan took him everywhere with her: to acupuncture appointments, to Lowe's, to friends' houses, on walks in the woods to explore. I followed his exploits via e-mails and photos. He had everything a puppy could desire and beyond. Everything, that is, except other puppies to interact with. His young mother, Vita, an intense West German import, wasn't a mentor. Her idea of mothering Solo was to nurse him frenetically and then race away like Road Runner from Wile E. Coyote, leaving him "in a cloud of dust." So Solo's great-aunt Cora, with her fawn-colored coat and sweet face, her impish sense of humor and tolerance for unusual puppies (because she had been one herself), took over the task of raising him. It is always thus in extended families, and some are the better for it. Solo interested and amused Cora. She taught him her love of toys and games, and he got away with everything. In one picture, Solo is walking across Cora's reclined body, carrying his favorite stuffed duck, leaving dents in her plush coat.

Solo was no longer a squashy mole; I could now see that his head was going to be glorious. Part of that big block of gorgeous was dedicated to his olfactory system. Even at a fast run toward Joan, he often screeched to a halt, nostrils flaring at some wayward scent. "His nose rules," Joan said. That wasn't welcome news. Megan, because of her hunting lineage, froze at the sight of a bird, cat, or squirrel, every synapse alight and devoted to that one task. I had planned for a dog who would focus only on me. I knew it was going to take a year or so to get him up to speed, but I'd always watched with a touch of scorn as obedience handlers with flop-jowled bassets and beagles had to plead with their dogs to raise their snorkeling, scent-mesmerized noses off the ground and pay attention to them.

It was mid-May 2004 and already hitting the eighties in Ohio, a preview of ever hotter summers to come, when we drove the 450 miles from North Carolina to meet and pick up Solo. He was lying alone in an open cage on the front lawn when we arrived, a still life in red and black, one paw tucked under his chest, relaxed, surveying his domain. He was already past the brief cute phase that shepherd pups have when their ears are soft and floppy and their noses don't yet look like shark snouts. Solo greeted us briefly, sniffed us, ignored us. He ran around grabbing at toys, pushing them at various adult shepherds. He had nerves of steel. He was full of himself. He made me slightly nervous. Joan had arranged a lovely dog-and-people party to launch us back down the road to North Carolina. Solo ran, growled, and leaped during the entire event. He said farewell to his dignified father, Quando, by grabbing and holding on to his bright gold scruff. He finally had to drop off when Quando looked down his considerable Roman nose and backed up slightly.

We gathered up Solo and his precious toys and drove down the country road, back to North Carolina. In the rear seat, locked in a travel crate for his safety and ours, lay our furry future. I don't remember much about that long drive except that it was hot and Solo was a perfectly equitable traveler, happy to hop out of the car, wag, do his business, and clamber back into the crate like a miniature adult shepherd. I started feeling better about him.

"Oh, my," said our friend Barb Smalley, who arrived that night to witness the homecoming. She watched as nine-week-old Solo leaped on Megan, bit her ears, and tried to hump her. "He's quite something, isn't he?" David and I were exhausted. Solo wasn't. Megan was drooling and panting in distress. I already looked like a junkie from my efforts to intervene: My arms had black-and-blue puncture marks where Solo had swung back on me in a frenzy.

He spent his first night with us whining and growling, methodically chewing through an inadequate and expensive fabric show crate. Solo wanted to continue his evening. I cried in David's arms. I wanted our whimsical, gentle Zev back. His worst sin had been to take a bar of soap from the shower and place it carefully on the bathroom floor with one faint canine tooth mark.

"I don't like him," I wailed above Solo's whines. I saw a grim future, a German shepherd roaring through our house and marriage, leaving shards of pottery and anger.

David firmly and kindly said exactly the wrong thing: "We'll just return him." My sobs redoubled. He later claimed he said it only to kick-start me out of my depression.

In the morning, I woke up and armed myself, grimly strapping on a belly pack loaded with greasy liver treats. I picked up a plastic-and-metal clicker that would make a metallic "tock" to mark the exact behavior I wanted. The little bastard—I would shape and mold him with clicks and patience and treats until he was dog putty. Or at least until he stopped trying to hump Megan. I had already given up on the dog-who-would-sleep-in-my-study-while-I-wrote fantasy.

David and I both fell hard for him. I fell harder because I always yo-yo further than David. By midday, I was laughing and infatuated. Solo was a maniacal clown, a Harpo Marx. Funny and charming. At least around David and me. He thought we were the cat's pajamas. He told us all about it: mewling, growling, barking, yowling, whimpering. He was operatic in range and expression. I'd never heard that kind of variety except on *National Geographic* specials about the wild dogs of Africa. Solo would stare at us, make a wolflike "rooo" sound, then try a gymnastic move to see our reaction. He found toys and leaped on them and brought them to us and dropped them and backed up. He started to learn their names. He played and played and played. With us. Not with Megan. He tried to bite us and then collapsed in our laps and fell asleep, twitching. When he woke, he fixed a gimlet eye on us. Game on. If he wasn't sleeping, he was watching us, waiting for the Next Big Thing.

On night two, I didn't sob. Partly because I was exhausted, partly because I was realizing that we had something peculiar and exceptional on our hands. Solo was diverting me from despair. David, who valued intelligence above almost everything, was smug but tried to suppress it. We had, he realized, the smartest dog he had ever known.

Smart didn't mean peaceable. Megan remained in shock. She stared at us without seeing, the whites showing at the edges of her large brown eyes. To handicap Solo a bit more, I soaked her fringed ears and tail in bitter-apple spray so he was less tempted to swing from them. That second night, she used her entire sticky body like a caterpillar's to hunch her foam bed as far as possible from Solo's crate in the bedroom, inch by inch. I. Do. Not. Like. That. Puppy.

Solo didn't care. Megan was just a dog. Dogs weren't his people. Solo had no litter to miss. We had no need to put a clock in the crate to mimic the sound of siblings' beating hearts. He slept through the night. He was at home alone.

Over the next two days, David and I tried to teach Solo the international language of ouch, something he'd missed out on with no other puppies for interaction. Joan had taught him, of course, but Solo found it convenient to forget, with new hands to bite. We screeched every time a sharp puppy tooth hit skin. Solo didn't relate, though he did cock his head when he heard our howls. Since he'd never experienced pain in exchange for his excesses, because of his kind and patient adult shepherds, he had no idea what it meant to cause it.

On day four, Megan stopped drooling and looking betrayed. She gave Solo a brief, queenly play bow. Permission to engage. She started to teach him a few basic commands to quell his most brutish tendencies. No more mounting. No humping at all. No more standing over her when she was lying down. No more massive puppy paw placed on the shoulder. She would move sideways a fraction so that Solo's leaps ended with him splayed on the floor rather than on her gorgeous setter body. She glanced over at us and opened her mouth slightly to show her small white teeth, smiling. Within a couple of hours, her tail returned to its former high-flying flag position, though her long silky feathers had gaps torn in them. For the first time since we had brought her home from Oregon three years before, David and I were in awe of Megan. Our space cadet had disappeared. We watched her, trying to learn from her engagement and disengagement, her covert and canny manipulation of this emotionally stunted puppy. We wanted to know what Megan knew.

We also watched Solo. I began to understand what Joan had meant when she repeatedly mentioned "scent drive" in her e-mails. David went out to work in the garden, and I left the house with Solo five minutes after that, wanting to avoid accidents in the house. He wasn't interested in peeing; instead, he put his nose just above the warm stones of our courtyard and started moving fast. Then he was on our crabgrass, skimming the ground, his pointed ears making his head look like a shark snout. He didn't lift his head until it butted hard against the legs of a startled David working behind the greenhouse. Solo had tracked him a hundred feet from the house, around two bends, on three different surfaces. Solo's whole body wiggled in pleasure, and he bit David's jeans happily until David "ouched" him. Solo had done his first short track. I had a new command that he loved: "Go find David."

We'd had Solo just a few weeks when my father and stepmother, Angie, came to visit from Oregon. Dad, his skin increasingly loose across the big bones of his hands, sat blissfully stroking Megan. He looked tired, but then he had looked tired most of his life. I tried to keep Solo—the inverse of what Dad liked in a dog personality—

exhausted and as far away from the three of them as possible. We plied Dad with David's home-baked bread, black tea at three P.M., a Scotch on the rocks at five P.M., and long political conversations. Dad was proud that I had settled down with my own Ph.D. and a fine, loving husband. It had taken me longer than it should have, over forty years. But—and I know that Dad thought this with the noblest of intentions, because we talked about it—he had great hopes that I could get down to the business of deep academic thinking. Now that smart, funny, dependable David was in my life, Dad no longer had to worry about what the vampires of loneliness might do to me. Now that I was an academic, my father no longer had to worry about my taking risks, as I had when I was a newspaper reporter, covering chemical leaks, natural disasters, and criminal trials. Dad had raised us in Corvallis, Oregon, recently judged the safest city in the United States: no earthquakes, no hurricanes, no twisters, no extreme weather. Nothing. He was pleased that my life was almost as predictable as when we had lived there, except for a hurricane every now and again.

I must have turned my back. That was when Solo leaped. Blood welled on the back of Dad's hand from one of his huge blue veins. He dismissed it with a shrug. Even if he was a German shepherd, Solo was just a puppy. I put Solo in his crate with a goat knuckle bone to gnaw. The weeklong visit was ending. Dad and I walked slowly one final time around the yard before he and Angie left to fly back across the country. The two of us looked at the new blueberry bushes, strains developed for North Carolina temperatures and humidity. We admired the male cardinals that dropped like red explosions from the willow oaks to the ground with their distinct cries of "Chew! Chew! Chew!" We talked about how good the future looked for both of us. Dad's undiagnosed cancer was probably well on its way to metastasis that June.

· · ·

My simple commands to sit or heel or "settle" didn't interest Solo. His crazy energy reigned. For him, it wasn't enough to walk into the mudroom, lie down, and wait for dinner. He had to launch himself, twist in midair, plié, and crouch like a gargoyle, lips pulled back in a grin. By braking hard a few feet before the door, he attempted to slide and somersault. He had a great sense of humor.

Solo adored David and me—and even Megan. He was an unpredictable sociopath with other dogs. Solo thought they were hostile aliens. Especially shepherds or other dogs with pointy ears. He developed a reputation early: The moment he smelled an unfamiliar dog, he bristled and growled. That made veterinarians' offices a challenge. One vet put in her notes that I was on the way to having a mean dog. Solo was ten weeks old. I quit her. Another recommended expensive acupuncture and homeopathy. I quit her. Puppy classes weren't a good fit. He would walk into a training center already barking and growling, hackles raised. The same obedience handlers who used to smile when they saw Zev arrive for training snatched up their shelties and schnauzers at the sight of Solo. AKC obedience trainers with decades of companion-dog and utility-dog titles to their credit strategized with me. Perhaps Solo needed a new kind of halter, maybe a Gentle Leader, to guide his wayward muzzle through the Sturm und Drang of his life. One trainer with whom Zev and I had worked for years suggested that I needed to discipline Solo severely for his behavior. I quit her. I was getting good at quitting.

I'd never had a dog-aggressive dog. It was like a scarlet A, confirming what many people already knew: German shepherds were dangerous. I dove into research. I ordered expensive videos and books on canine aggression. My fearful anticipation of Solo's reactions would travel down his leash and right into his limbic system, making the overall effect exponential, confirming what he already knew: Dogs meant trouble. That made Solo even more trouble. My relationships with other dog owners began to suffer.

A trainer accustomed to cheerful Labrador retrievers looked at me with dismay and shrugged after Solo roared and leaped on a short-haired pointer who had bounced over to greet him when Solo was on a down-stay in an obedience class. "For God's sake," I snarled, mostly to myself, "keep your dog on a leash and under control." I was at the end of my own leash. I was starting to get people-aggressive. I left the class early and e-mailed Joan, who sent back reams of wise advice. I cried again that night in David's arms—angry at Solo, at the stupid pointer owner, at myself.

Now it was much worse. I wasn't simply invested in the idea of this pup: I loved him. He was my dog and my responsibility, lying quietly in the backseat on the long drives home from failed puppy classes. I was failing him miserably, seesawing between training systems, avoiding places I knew other dogs might be.

None of Joan's many shepherds had exhibited Solo's behavior around other dogs. Later, she told me that when Solo was born, she had worried. Knowing what we might be facing, she had tried early on to find a litter to put "HRH" with, to no avail. No one in the immediate vicinity had a litter close to his age.

"Quite honestly," Joan e-mailed, "I am now convinced from years of training, and now Solo, that the majority of puppies missing a complete litter experience just don't learn how to handle the nuances of a variety of dog interaction. They don't learn the give and take."

People kept asking me, after watching Solo growl and leap: "You named him after Han Solo, right?" No. Absolutely not. I had never liked *Star Wars* much, and I didn't like the character, even if Harrison Ford had played him. Yet the description fit Solo to a T. Charismatic. Selfish. Brash. A talented, reckless misfit.

A loner.

2

Death and the Dog

Although they have been cherished for their good qualities—hunter, guard, herder, friend, worker—the inverse dog is the spoiler of human graves and eater of corpses, the keeper of hell's gates. . . .

—Paul Shepard, *The Others: How Animals Made Us Human,* 1997

Two months after Solo's arrival, I found myself in Nancy Hook's backyard in Zebulon, perched on the edge of an aluminum folding chair. Nancy slumped back in her sturdy canvas chair, her hand wrapped around a foam beer insulator wrapped around a Gatorade. She was mellow except for the warning she occasionally gave the dogs quarreling in the kennels next to the yard: "Don't make me come over there." They stopped. It was mid-July and too hot to fight, in any case.

Japanese beetles clattered past. Tent caterpillars had wrapped up and skeletonized half the leaves of the huge pecan tree we sat under.

I knew Nancy from when I'd taken Zev to her parking-lot obedience class some years before. She had been welcoming and kind to both of us, though not particularly interested in Zev. He had been so mild-mannered that he tended to disappear in a dog crowd.

I hadn't seen Nancy much since, but I started to remember as I pulled into the drive and read the black bumper sticker on her pickup: "Gut Deer?" modeled after the "Got Milk?" campaign. Her hair was still copper, her dark chestnut eyes still surrounded by smile wrinkles. She wore camouflage pants.

I had e-mailed Nancy in desperation, remembering her sense of humor and practicality. I needed both. Sure, she said, come on out to Camp Hook. Bring the dog. She was competent and relaxed; I was edgy and talkative. Solo, more obnoxious than any four-month-old German shepherd should be, was hackled and humpbacked, wild-eyed and ungainly. From time to time, he surged toward the kennels, a dark hybrid of colt and Tasmanian devil. He would snarl and bounce off the cyclone fence. I bounced off the lawn chair, wrestling Bil-Jac dog treats out of my fanny pack, trying to distract him and minimize the behavior that Nancy was witnessing. "Solo? Solo? Watch me! Gooood dog!" I funneled liver into his mouth.

"Stop chattering at him," Nancy said. "And stop giving him so many treats. You're making him into a wuss." My hand froze in mid-dive. "He's just a jackass," she said. "What do you want to do with him?"

And with that simple question, my weird dog world started righting itself. By "What do you want to do with him?" Nancy didn't mean endless rounds of dog counseling and dog tranqs, creating a sedated and submissive shepherd who needed an occasional cautionary Dog Whisperer "hisst" with an index finger held up to keep him in line. Nor did she mean that I could click-and-treat this dog into executing perfect obedience routines. That didn't work with him; besides, I was bored with the obedience ring. Nor did she mean that Solo was capable of becoming the quintessential park dog who would allow me to sit on a bench with other tranquil owners, gossiping, watching our dogs romp and bark into the sunset.

She meant: What would you like this dog to do?

I had no idea. I wanted him to be so busy that he didn't have time to do what he was doing in front of Nancy. I wanted him to have a job, if possible. Not a pretend job that would simply exercise out his little heart of darkness. Probably not a job as a therapy dog in a nursing home, because of his rhino ways. I wanted his work to have meaning, as I was constantly struggling to find meaning in my own work.

Nancy didn't indulge my angst for long. "Stop thinking so much," she said. "That's part of your problem."

She ordered me to leave Solo alone. I pulled my hands away from the greasy treat bag and put them at my sides. I turned my gaze away from Solo's evilness. Within a couple of minutes, he came over and flopped in the shade. Being bad wasn't as interesting if I weren't reacting.

Nancy and I talked, running down my options. She taught everything from housebreaking to bite-breaking to obedience and trailing. Training Solo for search and rescue wasn't ideal. I couldn't leave students waiting in the classroom for a lecture on feminist essentialism because I was running off to search for a lost three-year-old who was in fact playing with Transformers at the next-door neighbors' house. Nor could I count on my own body being the ultimate fitness machine, capable of running for miles after a dog tracking in thick underbrush; I might end up asthmatic, shambling, sciatic nerves aflame, eyeglasses either fogged or smashed. Lost people needed better odds than I offered.

That made sense to Nancy. Besides, she had become less enamored of search-and-rescue team politics over the years. She described them in ways that made them sound similar to my English department, without the Victorian charm. More issues emerged. I didn't want to wear search gear that would make me look like a Girl Scout. Then there was the idea of a team. I could collaborate, but I couldn't really relate to the cheery phrase "Remember, there is no 'I' in team." It didn't suit Solo, either. Better that he didn't constantly have to deal with the hurly-burly of dog society. Sending him out to track alongside several self-assured search dogs? They wouldn't put up with his nastiness. They'd reduce him to tufts of black-and-red fur spread over the trail.

There was one way around all of the scheduling problems, my team-player problems, and Solo's psycho-puppy problems. Nancy was pleased with herself for coming up with it: "a cadaver dog."

I didn't know exactly what Nancy meant, but I could guess. Dead dog. I'm good at putting words together and knowing what they mean. It's what I do for a living.

It's ideal, she told me. The dead will wait. In the meantime, they emit scent. With a few frozen exceptions, more and more scent over time. And cadaver dogs and their handlers work mostly by themselves, in methodical search grids, not alongside other dogs and handlers. The dog's job is both simple and complex: to go to where the scent is the strongest and tell the handler it's there. It's work that needs to be done. Families and law enforcement, mostly, although not always, want bodies found. Besides, she told me, beaming, her smile lines in full evidence, "It's a ton of fun. You'll love it!"

Nancy avoided mentioning that my salmon-colored linen pants were probably not the ideal thing to wear on searches.

At the end of our session, she sent me and Solo off down the road. I was sweaty, I reeked of liver treats, and I was filled with inexplicable happiness about those who go missing for a long time. Solo, exhausted, slept soundly in the backseat, although his outsize feet continued to twitch, pedaling air-conditioning instead of cyclone fence.

Nancy, knowing my compulsive habits, expressly forbade me to read about training dogs on cadaver scent. I would screw up Solo's training by reading too many theories too soon. She had two exceptions: Bill Syrotuck's *Scent and the Scenting Dog* and Andy Rebmann's *Cadaver Dog Handbook*. I ordered the two books. Then, because waiting isn't my forte, I sneaked onto the web to learn the basics of death and dogs.

. . .

No house would stand firmly founded for me on the Ahura-created earth were there not my herd dog or house dog.

—Ahura Mazda, Zoroastrian god

In 2012, archaeologists in the Czech Republic published their discovery of three skulls of what appeared to be domesticated dogs, shorter of snout and broader of braincase than their wolf cousins. One of the skulls, 31,500 years old, had a flat bone fragment, probably that of a mammoth, inserted in its jaws. It was so purposeful and evocative that the archaeologists couldn't help speculating: Was that bone part of a funerary rite, appeasing the spirit of the animal, inviting it to come back, or encouraging it to accompany deceased people?

The speculation wasn't much of a stretch. For all that dogs seem to lurk on the edge of civilization, we've also let them in and granted them special status. For thousands of years and in numerous religions, the living have depended on canines to help guide the dead—to get us from here to there, wherever there is. Few myths have such worldwide resonance. One can see the temptation of assigning dogs this task: They appear custom-designed for it. Dogs howl at the moon, warning us that death is just over the horizon. They can hear and smell, growl and hackle, warning us of specters that our dull senses miss.

They also like to eat things. Even us, given the opportunity. Dead people aren't so different from other dead animals. We're protein. Given an opportunity, dead people get smelly. We become deeply attractive not only to bottle flies but to more developed animals. Like dogs.

Part of the religious connection of death and dogs no doubt comes from a ritualized spin on the grim but useful reality that dogs and other canids, like jackals, scavenge. People witnessed that behavior—done with joy and impunity—and came to the obvious conclusion that dogs and their close relatives must be powerful, immune to the demons of death surrounding bodies. That made canids useful beyond the simple housekeeping function of getting rid of bodies. So in ancient Egypt, in a simultaneously pragmatic and religious switch, the jackal-dog became a god. Anubis, friend of the dead, was a protector, not a predator, of the deceased in their tombs.

While artwork and accounts of Anubis are plentiful, we have only one or two nineteenth-century accounts about how the ancient Bactrians (in what is now Afghanistan) and the Hircanians (then part of the Persian empire) handled this canid propensity. Those accounts note that the Bactrians used dogs called *canes sepulchrales*. The dogs had a specific job description: to eat the dead. In exchange, they received the greatest care and attention, "for it was deemed proper that the souls of the deceased should have strong and lusty frames to dwell in." It was a pretty nice deal for the deceased, who then got to hang out in a mobile furry coffin. The limited history doesn't note what happened after the dog died.

In Persia, the Zoroastrians made canids' roles more layered and central to mortuary rites. Like the Egyptians and Bactrians, they clearly decided to make the best of canids' tendency to love smelly protein. Zoroastrians were already using working dogs as a central part of their ancestors' nomadic herding existence. Mary Boyce, considered the greatest scholar of ancient Iran, wrote that "mortal dogs receive a striking degree of attention" in Zoroastrian holy texts. They likened the dog to fire, both protective and destructive. "It seems probable that this power came to be attributed to the dog because dogs are the animals always referred to in the Avesta as devouring corpses," Boyce wrote.

It takes some real mojo for dogs to do that and not be harmed by Nasu, the demon that brings putrefaction. The funerary rite in Zoroastrianism was called the *sagdid,* "seen by the dog." It took a special kind of dog for this work. A kind of German shepherd-like dog. The ideal *sagdid* dog was to be at least four months old and male, "brownish-golden" with "four eyes"—perhaps not unlike rust-and-black Solo, with twitchy black spots of fur over his eyes. One of the small cast-metal art objects in the Tehran museum looks like a stocky German shepherd, although German shepherds didn't exist then. The dog could be white with tawny ears, probably not unlike what we see in the Canaan dog of Israel, an ancient herding breed still in existence, or one of the guard breeds of that area.

The dogs chosen for *sagdig* got paid for their work. Zoroastrians knew their dog training. Three pieces of bread were placed on the corpse to induce the dog to approach, gaze steadily on the body, and drive Nasu away. That would be exactly how I started training Solo to both recognize and happily approach the scent of human death—only I used liver treats and then toys, rather than bread, to draw him in.

The work of dogs didn't end with *sagdid.* After the four-eyed dog was done with his job, corpse bearers took the body away, and the village dogs and vultures followed and feasted.

Zoroastrian dogs—from the herders to the hunters to the house dogs and the village dogs—had a pretty good deal: They got especially well fed when people died, and not just by getting a bit of bread or helping dispose of the bodies. They were given a whole egg and portions of the food offerings for the dead. When Zoroastrian house dogs died, they got extra-special treatment: Boyce noted, "Until the mid-20th century when a house dog died, its body was wrapped in an old sacred shirt tied with a sacred girdle, and was carried to a barren place, and brief rituals were solemnized for its spirit."

All the rituals sounded lovely, especially the one for the house dog. It was a step up from what we did with Zev after he died: We got his ashes in a hard plastic canister from the vet. The canister came with a burgundy velvet sack with a small rainbow bridge embroidered on it. The canister still sits in my great-grandfather's oak secretary. I don't know what we're waiting for. We should probably carry his ashes to a barren place.

. . .

> Achilles' wrath, to Greece the direful spring
> Of woes unnumber'd, heavenly goddess, sing!
> That wrath which hurl'd to Pluto's gloomy reign
> The souls of mighty chiefs untimely slain;
> Whose limbs unburied on the naked shore,
> Devouring dogs and hungry vultures tore.

—Homer, the *Iliad*

Westerners haven't been nearly as kind to canids as the Zoroastrians were, although we should have been deeply grateful to the female wolf who fed and raised Romulus and Remus so they could found Rome, our version of civilization. In the Western world, we balk at the notion of including dogs in our religious life. We're genuinely repulsed by the idea of dogs eating people. Homer used dogs' attraction to bodies to open the *Iliad,* the perfect frame for horror and chaos.

The large, evil, and almost always dark dog lurks on the edge of Western civilization: Hecate, the Helenic goddess of ghosts and witchcraft, had a black bitch familiar at her side. Greeks used to sacrifice black puppies to Hecate; dogs were a favorite sacrifice in a number of religions. Cerberus, the three-headed monster dog, let new spirits enter the realm of the dead, though no one could leave. Gamr, a bloodstained watchdog of Norse mythology who looks a lot like a German shepherd, guarded the gate to the underworld where evilgoers went. The Cŵn Annwn, Welsh spectral dogs, foretold death.

At least those polytheistic religions allowed dogs to play a variety of roles—they might devour bodies, but they were guardians and guides, too. Homer may have opened the *Iliad* with ravening dogs, but in the *Odyssey,* he used his hero's dying dog as a symbol of faithfulness: Argos was the only being to recognize Odysseus when he returned from his travels after a twenty-year absence. Those diverse dog roles didn't carry over to monotheism. Historian Sophia Menache of the University of Haifa posits that Jewish, Christian, and Muslim religions were threatened by dogs and the "warm ties" people had to them. Dogs had a central role in agrarian life; they reminded monotheists of the ever-present competition of animal-worshipping cults. So when we ask questions about organized religion and we ask, "Yes, but was it good for the dogs?" the answer is no. The New Testament's thirty-two mentions of dogs are mostly negative. Though the antipathies and insecurities of the three religions have softened and shifted somewhat in intervening centuries, dogs get short shrift in many Muslim countries, and some Christians want dominion over the natural world.

Even today, in a secular Western world, we remain oddly fascinated by the role that dogs play in death. Scamp, a schnauzer at an Ohio nursing home, got wide news coverage in 2007 for his habit of barking and pacing near patients' rooms when they were about to die. He had "eerily" raised the alarm for forty deaths in three years, the director of nursing told *Inside Edition*. Far from shunning him, the patients adored him. "It's not like he's a grim reaper," director of nursing Adeline Baker told the reporter. "It's kind of comforting to know that maybe at the end of our lives, if we don't have family members, there will be somebody there to be with us."

Perhaps Scamp was a comfort because he wasn't large and black but small and gray with quizzical eyebrows. The darker side of our superstitions has also survived: More big black dogs are reportedly euthanized in U.S. shelters than any other size or shade.

In modern times, we have updated and sanitized the Homeric language of "devouring dogs." Forensic scientists now call it "canine predation." Despite having a name for the phenomenon, we tend to keep an uneasy distance. Yet, a few years ago, children streamed into our local science museum for one of its most popular shows ever, on bugs and death. Shows like *CSI* and *Bones* have made us surprisingly comfortable with maggots, and what they tell us about the stages of death, and have contributed to the popularity of an entire discipline: forensic entomology. Scientists know a fair amount about bear activity; they know less about dog activity. Yet the handful of available studies show that dogs and their coyote cousins account for much of the scavenging on human remains.

The media appear to know a lot about dogs finding the dead. The problem is that stories are scattered everywhere, hundreds and thousands that all have the same innocuous story line: A person walking a dog finds a body. I am convinced that an analysis would show that untrained dogs out for walks or roaming neighborhoods find more bodies than trained dogs do. It's a simple question of acknowledging the millions of dog noses out there working unpaid overtime.

Depending on your perspective, it's either good or bad to let your dog roam off-lead, but let's face it: Dogs on leads don't find bodies nearly as often. Generally, finding a body is a good thing, although the dogs' owners and walkers are never thrilled when it happens.

Ollie, a golden retriever, was in Hollywood Hills on an unleashed walk in January 2012. A professional dog walker and her mother had eight other dogs with them. Ollie dashed into the underbrush and started playing enthusiastically with a plastic bag: "He was digging, digging, digging, barking," the dog walker, Lauren Kornberg, told the local radio station. Ollie shredded the bag and came away with something big and round in his mouth. He dropped it, and it rolled into a ravine. Kornberg admitted that it was her mother, "a responsible adult," who went to investigate—and found the head that Ollie had dropped.

A four-year-old black Labrador named Fish brought a decaying human arm into the front yard of his Mission, Texas, home in August 2011. Police were able to get the hand and arm bone before they disappeared down Fish's throat. The dog's adult owner was traumatized. Not so his eight-year-old daughter, who chatted with the television reporter. Their dog, she said, likes to visit the neighbors' chicken coop as well: "Fish gets everything. He brought eggs on Easter."

I understand her father's repulsion. I wouldn't accept Fish's gift of eggs.

When dogs become, in Paul Shepard's term, "the spoiler of human graves," it's a reminder of how we tend to deal with human bodies. We Westerners tuck them away fairly quickly. Dogs like Fish remind us of the disorder and chaos inherent when there's an arm or hand lying around where a dog can find it. We prefer hands either made into sterile ash or nicely preserved with formaldehyde and gently crossed over the body in a coffin. On the flip side, turnabout should be fair play. Both historically and in current practice across the world, people eat dogs without much compunction. There's good evidence that dogs were and are raised for meat; they were the first agricultural animal in a number of societies, and they remain so in some today.

None of my early research on cadaver dogs grossed me out. I realized there was a difference between reading about it and coming face-to-face with it, but abstractly, the idea of cadaver-dog work didn't offend my sensibilities. It made me happy. Perhaps my childhood in the woods and fields, growing up with fishing and hunting and gutting and plucking and skinning, was a factor in my sanguinity. Or the fact that I had taken care of my paralyzed mother and worked in nursing homes for years. Perhaps it was because my father was a biologist who taught me to look at dead organisms with a disinterested but not un-interested gaze.

Cadaver-dog work seemed straightforward to me. As one medical examiner and early cadaver-dog trainer, Edward David, noted with great cheer, "love of the putrid" is inherent in canines. So why not take that love and channel it toward something more socially useful than rolling in dead squirrel?

Why not take that love and see whether it might be used not to increase the chaos but to restore, even if only slightly, a sense of order?

3

Nose Knowledge

There are seventy-five perfumes, which it is very necessary that a criminal expert should be able to distinguish from each other, and cases have
more than once within my own experience depended upon their prompt
recognition.

—Sherlock Holmes, *The Hound of the Baskervilles,* 1902

These days, when I watch a good dog work scent, I can see him trace
its passage in the air until he's drawn a clear picture with his nose. An
experienced dog can illustrate the difference between scent that has

lifted in the heat of the day, settled down in the ridges of rough grass, or been pulled hard toward the rushing water of a creek.

I work scent also, though I don't run as fast or as hard. I can recognize urine in the muggy concrete stairwell of a parking garage, mildew liberated from under sheaves of rotted leaves, and the fishy musk of a German shepherd after a swim in the Eno River in August.

I knew, even before I started researching dogs' noses in a sustained manner, that they were much better than humans'. Solo's was much better than mine. David could use a pair of scissors on a package of vacuum-sealed meat in the kitchen, and the silent exhalation of bloody air would wake Solo and bring him running from across the room. Conversely, it seemed lazy to concede that dogs are the masters of the domain of olfaction before I'd done any scientific fact-checking. Sure, Solo has a bigger nose than I do, but size isn't everything.

So what's the truth about dogs' sense of smell? I don't want to keep the reader in suspense: The truth is, we don't know that much about it. As I began researching, I noticed wildly fluctuating figures in both the sentimental and scientific dog literature: The dog's nose was either ten times or a hundred times or a thousand times or tens of thousands of times better than the human nose. Those figures raised doubts. If scientists, or people who play them on YouTube, or your basic dog lover had stuck with one false figure—say, that dogs' noses are a thousand times better than ours—or if they had stated the figures with less certainty and more modesty, I might not have become suspicious. Given the variation, I wondered, how much better are dogs' noses, really? And if they are better, what are they better *for,* exactly? Sniffing dog pee?

In tracking rumors, as with tracking most things, it's good to start at their genesis. There's a growing body of scientific evidence suggesting that not only have the nose and its receptors been important to the survival of creatures for at least hundreds of millions of years; they may also have been a key evolutionary force driving the growing intelligence of mammals.

In 2011, Texas paleontologists published their analysis of the skulls of pre-mammals living 190 million years ago. Their research shows that one of our pre-mammalian ancestors, the *Hadrocodium wui*—a shrewish beast with a skull smaller than a paper clip—didn't have the option of rejecting smells. It tremulously ventured forth to sniff for grubs and insects, probably at night so the diurnal dinosaurs didn't accidentally squash it. Its fur was important, its twitchy ears important, its vision important. But, the paleontologists argued in *Science,* it was the critter's olfaction system that was the most impressive thing in that bitty skull: "[It] differed from even its closest extinct relatives specifically in its degree of high-resolution olfaction, as it exploited a world of information dominated to an unprecedented degree by odors and scents." They theorized that these skulls showed the olfactory system played a major role in helping the mammalian brain evolve, apparently to the point that we humans—the most advanced of mammals—could turn our collective nose up at thinking too much about smells.

We should feel grateful to the tiny *Hadrocodium wui,* fossilized and preserved in China, for helping us understand the importance of a sense of smell. Yet there's still much about this sense that scientists don't know. It was only recently that researchers started to decode how our olfactory system actually works. Thanks to Linda Buck and Richard Axel, who won the Nobel Prize in 2004, we've learned that when a volatile aroma latches on to an odorant receptor, it fires an olfactory neuron. Bam. That is not the only theory out there, but it's the reigning one. You would think that researchers would be further along in understanding this particular sense, but amid the general devaluing of the sense of smell in the Western world, the chemical complexity of odors, and the complexity of the neural circuitry that underlies even the act of sniffing, they have a ways to go.

As for comparing one nose with another? Neurobiologists aren't that interested in setting up olfaction competitions between species. They don't care which species' nose is "the best." What does that mean, in any case? This kind of speculation is better left to Animal Planet's *Creature Countdowns,* featuring the "Top 10 Animal Troublemakers" and "Top 10 Animal Skills." The bloodhound made number nine on that last list, with a sense of smell so fine, according to the producers, that it is "up to a million times more sensitive than that of humans." I am not making this up. Animal Planet is.

Some species' noses do rise above the rest, literally and figuratively. I bet if we could train bears to track, their noses would confer real bragging rights. Biologists believe that grizzly bears generally have a much better sense of smell than any dog. Bears' noses get less press than dogs' for all the obvious reasons. If you ever get the chance to stare at a grizzly's nose close up, you can see what a stunning instrument it is—tilted up at the end, with huge flaring nostrils. If you are at exactly the right angle, you can see through that nostril space to the blue sky beyond. The grizzly bear can manipulate her nose like a flautist can flex her fingers. If you could peer inside her *horribilis* skull at the nasal cavity, you would see delicate structures that look like two huge morel mushrooms or rounded honeycombs planted side by side. Their job is to process scent. How well? Pretty darn well. For miles, it's estimated.

But maybe not for miles and miles and miles. Nonetheless, Animal Planet, some books, and a number of bear websites use the following phrase: "Some scientists say bears can smell carrion from up to 18 miles (29 kilometers) away." It's rather odd, I thought, that "some scientists" would all agree on such a specific bear mileage. The figure's genesis, I discovered, was one 1976 bear-conference paper noting that one radio-tracked bear traveled twenty-nine kilometers with some speed and ended up at a carcass. The ecologist, Frank Craighead, stipulated, "It was not determined just when and how the carcass was detected."

Nor are there good comparative studies about the capacities of one animal's nose versus another, said Larry Meyers of Auburn University College of Veterinary Medicine. Repeat falsehoods enough, though, and they don't just take on the patina of truth, they become its replacement. Bears don't care, and inventions about their olfactory capacity probably matter less because bear handlers don't end up testifying in criminal court about their abilities. Nor are bears' noses depended on for finding lost children alive.

Besides, there's just no great bear-nose grant money these days. The same isn't true for research on human noses. We're always interested in ourselves, and it's easier and safer to get human noses, rather than bear noses, into a laboratory. Even while we tend to denigrate it, the human sense of smell is getting more deserved attention. Humans can detect thousands of different odors. Even Linda Buck, Nobel olfactory biologist par excellence, hems and haws over some details: "It is estimated that humans, for example, can detect from 10,000 to over 100,000 different volatile compounds." That's a difference of a factor of ten, less than some of the variations I've seen used with dogs, but hardly a rounding error.

While smell isn't entirely a lost subject for research or practice, for most Westerners, it is a deeply underappreciated sense compared to vision. It wasn't always so. Smell used to be a critical tool for physicians. Hundreds of years before we started exclaiming over the miracle of dogs being able to detect diabetes or lung cancer, doctors were using their noses to do the same thing. "Evaluating effluvia" was considered a basic diagnostic skill: Sweat on a rubella patient smelled like "freshly plucked feathers"; life-threatening diabetic ketoacidosis made a patient's breath smell like "rotting apples"; a certain bacterial condition made the skin smell like "over-ripe Camembert." Now we leave that job to lab tests and litmus paper.

Vestiges of human scent skills do survive in pockets in the West. For instance, "odor mitigation" expert Larry Sunshine will fly into a city, tilt his head, open his nostrils, and identify specific foul odors in subways, including mildew and chemicals venting off plastic seats. Luca Turin, a perfume expert and biophysicist, can deconstruct a perfume down to its exact compounds and describe it in terms that make you want to laugh and weep and even buy that particular patchouli: "The smell was at once beguiling, salubrious, and toxic, and felt like a perfume composed for a fiercely intelligent librarian."

Research is starting to blossom in this world of human olfaction, and it's taking some interesting turns—ones that working-dog handlers can relate to. Certain scents turn humans into the kinds of tracking machines whose accomplishments seem to rival those of trained scent dogs. If chocolate is involved. Scientists took a group of Berkeley undergraduates, showed them a video of canine scent-tracking, and then took them out on a lawn that had been laced with essential oil of chocolate. Scientists gave the students blindfolds, thick gloves, knee and elbow pads, and let them loose, off-lead. Could they track the chocolate using their noses, crawling on their hands and knees? Absolutely. The track they followed looked like the zigzag of a dog's trail.

The Berkeley scientists loved the students for more than their noses. Unlike rats or dogs, they could be debriefed afterward. The students could even say which nostril they tended to depend on: "Humans are an appealing animal model for addressing such questions because they can follow task instructions and accurately report behavioral strategies," noted the article in *Nature Neuroscience*.

Fortunately, that's one aspect of olfaction that a few researchers are starting to focus on moving forward, both in humans and in dogs: the importance of training. Olfactory scientist Avery Gilbert has long argued that the trained human brain processes and understands scent differently. "I'd like to challenge my academic friends to stop giving random orders to college sophomores in the psychology labs," he wrote, "and start observing odor fluency where it happens naturally—in creative people actively engaged in smell."

In that respect, water is like chocolate. University of Pennsylvania researchers concluded in a study on people's ability to taste water that training, more than anything else, most enhances people's ability to identify smells. And during a repeat chocolate-tracking trial in Berkeley, students cut their tracking time by nearly two thirds. They learned to sniff faster. The students, taken out of the classroom and placed on grass, started to develop working noses.

Abalone and accelerants. Termites, truffles, and TNT. Crack cocaine and citrus cankers. Mildew, moths, melanoma. Peanut butter, pythons, and people. Spotted owls and spiny lobsters. Cows in heat. Gas leaks.

If there's a particular smell out there—illegal, endangered, delicious, destructive, invasive, or dangerous—handlers will try to train dogs to find it. The list of smells is lengthening daily.

Using hunting dogs who use their noses dates back many thousands of years. Human tracking and avalanche-rescue work originated centuries ago. But the explosion of tasks we've found for dogs in the past four decades often reflects our all-too-human tendencies toward violence, addiction, arson, and excess. We have entered the era of the sniffer dog.

Working-dog historians may argue about when that era began. The first handful of U.S. police-dog programs were developed at the turn of the twentieth century in New York City and New Jersey. Over the next few decades, units emerged in Connecticut, Berkeley, Pennsylvania, and Detroit. Fewer than twenty programs existed up through the early 1950s. After that, police K9 units reproduced rapidly.

Much of the expansion coincided with the military's experimental animal science that began in the mid-1960s and lasted into the early 1970s, during the Vietnam War, when researchers realized that sentry and scout dogs could be trained for additional sniffing tasks: finding punji pits and trip wires, not to mention illicit drugs in soldiers' barracks in Vietnam. On the domestic front, the military-research scientists speculated that dogs might be used to find bombs and weapons. Hijackings, bombings, and assassinations were on the rise in the United States: what one researcher called "the ills of the 1960s."

It wasn't just in law enforcement that the uses of dogs' noses started multiplying. Seattle area trainer and handler Marcia Koenig, one of the early volunteer search-dog team members, tracks the history of volunteer search-dog teams to 1962, when the German Shepherd Dog Club of Washington State formed the first air-scenting search-dog group. One of its founders, Bill Syrotuck, wrote the clear, concise book *Scent and the Scenting Dog* (one of the books Nancy Hook had permitted me to read). Today volunteer groups deploy to search in wilderness areas, in avalanches, on water, and in disasters. Marcia Koenig estimates that the U.S. has more than five hundred volunteer teams.

While working dogs were off and running by the mid-1960s, it's best not to feel too sentimental about what that growth represents. Each time a dog accomplishes a particular task for humans isn't automatically a moment for celebration. Dogs may have co-evolved with us, but they don't have a lot of say in how we decide to use them, so the "co" gives a false impression of equity. The dog mostly tries to please us, using its "canid tool kit of flexible sociality, a good nose, and expertise in hunting," as John Bradshaw, Foundation Director of the Anthrozoology Institute at the University of Bristol, puts it.

Good working dogs have to move swiftly, hear acutely, smell well, and communicate clearly with their handlers—even bite on occasion. Since domestication, they've been used as adjuncts for the evil that people do, as well as the good, and sometimes both at once. They can be used to consolidate or pervert power in concrete ways. They can track a slave, a lost child, or a rapist without distinguishing. They can help suppress peaceful civil rights protesters or control an angry mob that's up to no good. People create the problems, and working dogs come along for the ride. Right now, as we engage in conflicts in the Middle East and South Asia, we use dogs to find bombs and control the groups fighting against us. When domestic boas and pythons get too big for Florida apartments, they are dumped in the Everglades, and we use dogs to locate them before the snakes decimate the native wildlife. We have huge prisons filled with contraband and cell phones, and we use dogs to find them.

Though we've been using working dogs for tens of thousands of years, academic researchers are just starting to catch up with what these dogs do and how. Characterizing and understanding dogs' olfactory and cognitive skills isn't best done in a lab with limited and usually untrained dog subjects. If working dogs are overrated in the popular imagination, they have been mostly underrated in science, although that is rapidly changing. Nonetheless, far too much scientific work on dog cognition and olfaction has been done on pets—dogs who don't use their brains and noses for a living.

Just as Berkeley scientists took undergraduates out of the psych lab and onto the grass to track, just as Avery Gilbert urged neurobiologists into the kitchen of a great chef or the lab of a master perfumer to understand how human experts process scent, some working-dog experts are urging psychologists and neurobiologists to start using working dogs as research subjects.

The first scientific problem-solving tests with dogs appear to be those of Edward Thorndike, who served as president of the American Psychological Association at the turn of the twentieth century. He created "puzzle boxes" (precursors of the Skinner box), put domestic dogs and cats in them, and waited for them to figure out how to escape. The performance of both species was disappointing. Thorndike's conclusion was that operant conditioning, rather than any independent cognitive function, was the only thing that helped the beasts escape. The conclusion held sway in the research community for over a century. "Based on Thorndike's experiments and others like them, scientists now believe that dogs have rather limited powers of reasoning, certainly inferior to those of chimpanzees (and even a few birds)," John Bradshaw wrote in 2011.

And so it goes, even as dogs' noses become more and more popular with the public. For instance, one group of studies showed that dogs ranked from "okay" to "fine" at distinguishing between fraternal and even identical twins' scent on gauze pads. When dogs were asked to choose the scent difference between identical twins living in the same house and eating the same food, however, they failed miserably. That study's conclusions were clear about the outer limits of dogs' scenting capacity.

These are exactly the kinds of studies that irritate cognitive psychologist William "Deak" Helton, at the University of Canterbury, New Zealand, an expert in the science of working dogs. Experiments like the twin-distinguishing studies, he said, are akin to putting undergraduate students on a flight simulator, studying their performance, and coming to conclusions about the capabilities of trained pilots. One doesn't need to go all the way to the vastly experienced Captain Chesley "Sully" Sullenberger safely landing a US Airways jet on the Hudson River to think Deak might have a point.

"More studies need to be done on highly skilled dogs," Deak said. "The problem, of course, is these dogs are already likely to be working and are too valuable." It's not that the research done with untrained dogs is invalid, he noted; it's that those dogs haven't had their noses—or their cognitive abilities—trained and developed.

The twin-distinction story ends well. A group of scientists and ethologists in Czechoslovakia, where much of the groundbreaking work on dog cognition has taken place, decided to do another study on the identical-twin scent-discrimination problem. These researchers used trained scent-detection dogs. Their 2011 study showed that well-trained German shepherds could easily and correctly distinguish between the scent profiles of identical twins, even those who lived in the same environment and ate the same food.

Properly trained and handled, dogs will find almost anything we ask them to. For many substances, they can find small amounts. A 2006 study (albeit with a tiny sample) showed that trained dogs could detect one to two parts per trillion of n-Amyl acetate, a banana-scented solvent. That's the equivalent of a drop of water in twenty Olympic-sized swimming pools.

It's not universally true that dogs can smell at much lower concentrations than humans do. Larry Myers, who has been doing research on sniffer dogs since 1982, thinks it's silly to try to quantify which dogs have the best noses, which species have the best noses, or even to compare human noses with dog noses. Yet he couldn't help doing a couple of quick, informal experiments comparing his lab workers' noses with dogs' noses on odors. Myers tested one group of workers and dogs on acetone, a cleaning solvent common around labs. "My lab workers could smell it at lower concentrations than the dogs could." But with eugonal, a carnation-ey, clove-y compound that Myers started using as a standard pretest compound before having dogs perform more elaborate tests, they showed a response at one millionth of the concentration that lab workers could.

The old dogs' tales don't end at levels of concentration. There's also the strict correlation between the number of dogs' smell receptor cells and their scenting ability. At first, as I read the nose literature, I ranked a German shepherd's nose below a bloodhound's but well above most other breeds'. Solo's nose on this spreadsheet was very good but not excellent. Much of the best-selling dog literature promotes the concept that the more smell receptor cells a dog has in its nose, the better its scenting ability: "For example, the Dachshund has around 125 million smell receptor cells, while a Fox terrier has 147 million and the German Shepherd has about 225 million." The bloodhound has all those breeds beat, with 300 million receptors.

Nancy Hook thinks such distinctions are silly. If she wanted to, she said scornfully to me, she could train her daughter's Chihuahua, Pip, to find bodies. Lindsay drove the point home with smug delight by pointing out that her Pip could get into small spaces, unlike Solo, who is huge and clumsy.

Nancy and Lindsay were right. I ran across a National Institute of Justice report that quoted Lester Shubin, then a program manager with the NIJ. He and another researcher, Nicholas Montanarelli, would go on to collaborate on a number of projects, but at that time, Nick was a project director for the U.S. Army's Land Warfare Laboratory in Maryland. He was a military researcher who had started thinking early on about the potential of working dogs. He and Shubin were early proponents of bomb-detection dogs in the mid-1960s, when skepticism ran high about any dog's abilities. The two men didn't consider just German shepherds or Labradors—they worked with poodles and other breeds. Unlike me, Montanarelli and Shubin had open minds, uncontaminated by the love of shepherd.

"We learned that basically any dog could find explosives or drugs, even very small dogs like Chihuahuas whose size could be an advantage," Shubin wrote in the NIJ report. "Who is going to look twice at someone in a fur coat carrying a dog? But that dog could smell a bomb as well as the German shepherd."

. . .

And when the bloodhound came to the chief market-town, he passed through the streets without taking any notice of any of the people there, and left not till he had gone to the house where the man he sought rested himself, and found him in an upper room, to the wonder of those that followed him.

—Robert Boyle, *Essays of the Strange Subtilty, Great Efficacy, Determinate Nature of Effluviums,* 1673

Enough of Chihuahuas and German shepherds. Consider the bloodhound. Surely there's no debate about that nose being the best in the business.

Of course there is. I didn't think much about the bloodhound before getting Solo. What little I knew about its history, restrictively framed by popular culture, made me uncomfortable. With all that loose skin and hulking bone and wrinkle, the bloodhound is a generous and expandable doggie container into which we can throw all the myths, contradictions, drawbacks—and yes, even the real wonders of the working dog. It's a pity that there's so much contradictory nonsense about bloodhounds, and that I briefly bought in to some of it: not the silly pictures of bloodhounds wearing Sherlock hats and smoking pipes but the more serious nonsense about the bloodhound nose being the purest and most advanced miracle of nature, or the claims of bloodhounds being able to follow four-month-old tracks and trail cars for miles down freeways.

Exaggerations about the bloodhound's nose distract from the truth about a fine single-purpose trailing dog. If you want to watch a working dog's nose do its business, there's no more beautiful sight than that of a good trailing bloodhound. On the other hand, I've also watched Belgian Malinois run wonderful trails, as well as Labradors, Plott hounds, one Weimaraner, and a bunch of mutts. And, of course, German shepherds. Chihuahuas are probably a stretch for trailing, at least over long distances. Let's face it: None of these breeds is as evocative as the bloodhound. Or as storied. In Shakespeare's *A Midsummer Night's Dream,* bloodhounds were an absolute vision: "So flew'd, so sanded, and their heads are hung with ears that sweep away the morning dew."

Terry Fleck, a law enforcement K9 expert who has had three German shepherds, believes that a major difference in reputation for trailing is simply that temporal edge: Bloodhounds have a few hundred years' head start over German shepherds and Malinois. "History is working against us."

It's a hound-fulfilling prophecy. People believe that bloodhounds are great trailers and trackers, that the breed is the only one capable of working old trails. Someone lost? Bring in the bloodhounds. Combine that with an enduring interest in the breed's tracking ability. Seventeenth-century philosopher and founder of modern chemistry Robert Boyle described with both scientific detachment and fascination witnessing a bloodhound following an aged trial track, four miles long, with great cross-contamination: "[T]he dog, without ever seeing the man he was to pursue, followed him by the scent . . . notwithstanding the multitude of market people that went along in the same way, and of travelers that had occasion to cross it."

U.S. courts have followed history's footsteps. Early on, courts required proof of purity of bloodhound stock before they would accept trailing evidence. As an Ohio court noted in 1896: "It is a matter of common knowledge, and therefore a matter of which courts will take notice, that the breed of dogs known as bloodhounds is possessed of a high degree of intelligence and acuteness of scent. . . ."

It's not just legal precedent: Habit and niche needs work against us. Practically speaking, a law enforcement handler is going to train a patrol dog to track the guy who broke into Pizza Hut an hour before. Spending weeks or years training the dog to intersect and then follow a two-day-old trail isn't cost-effective, although more than one dog trainer thinks a shepherd or Malinois would do fine with a two-day-old trail. Bloodhounds have an edge because no one expects them to pay attention to anything but their nose on a trail.

"Dogs of any breed with the proper temperament that are trained the way top bloodhounds are can probably do as well as the best bloodhound," wrote retired Maine state game warden Deborah Palman. "One advantage of bloodhounds and other single-purpose tracking dogs is that they are not taught to hang on every word or gesture of the handler. They learn to track and do that job alone, without the interference of having to know sit, down, and heel."

The "history is destiny" sentiment also backfires against the bloodhound. Put a dog that huge and ungainly on a pedestal, and it's bound to fall off. All the overstatements (some of which come from a few delusional bloodhound handlers) work against the professional bloodhound handler who has to contend with the myths. A truthful, talented handler sounds like he's badmouthing the breed or his own dog when he's just being honest. A bloodhound and handler can accomplish something amazing, such as following a long, cross-contaminated eight-day-old trail—which, in its own small way, is like landing a disabled jetliner safely on the Hudson River—yet people think, "Hey, no big deal." After all, bloodhounds do that all the time.

Roger Titus is a big fan of bloodhounds, which is as it should be, since he's had fourteen during his long career. He has run behind thousands during trainings, and he's vice president of the National Police Bloodhound Association, "dedicated to the advancement of the man-trailing bloodhound." The association celebrated its fiftieth anniversary in 2012. Roger, who is seventy and still running, traces the rise and slow fall of the bloodhound in law enforcement with some regret. Part of the bloodhound's decline as a law enforcement dog is simply a question of time and money: Everyone is a pragmatist these days. A dog who trails like nobody's business but doesn't do anything else for a small or medium-sized K9 unit isn't going to be a department's first budget choice. If a police department is going to buy a dog, Roger notes, a German or Dutch shepherd or a Malinois—who can protect its handler, find dope, do short tracks, and bite people—will be considered the ideal candidate. But multipurpose dogs aren't always wonderful at every single thing they do. There are only so many hours in a dog and handler's week to train all the different tasks. And even if a police department does decide to commit to a bloodhound, it takes a special kind of K9 cop to want one. Here's the most delicate way to put it: It takes a handler who doesn't think the dog is responsible for carrying the handler's masculinity credentials for him.

One glorious fall morning, I watched with Roger while a young female bloodhound bounded right past the human scent trail laid for her four hours before, dragging her equally inexperienced handler behind.

"That little pissant," Roger said. Her galumphing leaps took her right out of scent. "She's going to have to slow her little butt up and try to concentrate." It wasn't just the young dog who was the issue; it never is. "I believe the boy wants to be a patrol dog handler," Roger said in a mild and almost nonjudgmental tone.

That's a pity. Despite all the myths, I've fallen for bloodhounds. Not only for their noses, but for their slobber that flies in big goobers, the concurrent flopping echo from their cavernous jowls every time they shake their heads, the red haw under their eyes that's punishment for those heavy jowls, their oily kennel stench, their tendency to gaze at you with a kind of walleyed abstraction, and the way they delightedly snorkel up hot dogs or Vienna sausages as a reward at the end of a long trail, like big wet vacuum cleaners. Later in their training, good bloodhounds don't need hot dogs. Running a good trail is itself the reward, just as the act of herding rewards an experienced border collie. Bloodhounds' doleful baying from the backs of trailers and pickups, with metal crates amplifying the bass tones, sounds like the essence of North Carolina foothills in the morning. Just to get rid of another myth: They don't trail together and bay like fools, "opening up" on the trail. It's a physical impossibility to sniff scent and vocalize at the same time. Try talking or laughing and inhaling simultaneously. Bloodhounds make noise if they lose a trail and get frustrated; when they pick it up again, they fall silent.

I've also fallen in the love with the fact that bloodhound personalities are all over the place, as dog personalities should be—don't buy the AKC bloodhound lobbyists' line that all of them are slow, gentle, and mellow. Nonsense. How boring that would be. The dogs should be able to run hard and pull hard; if a dog is any good, her handler should be a poster child for rotator-cuff surgery. Some working bloodhounds have a decided edge. I don't walk up to an unfamiliar bloodhound anticipating automatic adoration any more than I do a German shepherd I don't know.

It's understandable that a few bloodhound lovers will try to sell the public on the notion that they are harmless goofballs. Liberal dog lovers like myself, even more than dog lovers generally, can twist themselves into knots trying to explain the Byzantine history of a breed and its uses and misuses. (Trust me, I know: I have a German shepherd.) So a certain genre of breed fan will spend futile time explaining how incredibly sweet a breed is, trotting out docile "ambassadors" to disprove a bully breed's reputation. Dogs sleep just fine. They aren't monitoring their position on the annual U.S. dog-bite index. Whether they have bitten someone, justly or unjustly, or tracked an innocent person who was then arrested, they don't lie awake at four A.M. stewing that their true purpose—whatever the heck that is—was perverted.

As with most working-dog tales, the history of the bloodhound is steeped in lore, tiny tributaries of breeding programs that petered out, and popular-culture portrayals: from the noble early hound in medieval France, Le Chien de Saint-Hubert, to the sleuth hounds (a Scottish term) used in Europe for centuries to track both game and people. In the United States, we have McGruff, the animated crime buster, on the positive side; and on the other, the vicious prison bloodhounds tracking Paul Newman's character in *Cool Hand Luke*. Walt Disney's model for Pluto came from his grim 1930 cartoon *The Chain Gang*, in which Mickey escapes from prison and a pair of prison bloodhounds track him. When they open their huge mouths to bay, their fangs are enormous. It must be noted that it was the era when Mickey Mouse was skinny, had disturbing teeth as well, and looked like a rat.

It is slavery, however, that casts the longest and most inaccurate shadow on the modern bloodhound. You can see why the breed's devotees try to deny the accuracy of that history. They are right to. The dogs who tracked and attacked throughout the South during slavery and the Civil War have little relationship to the bloodhound in the United States today.

During slavery in the southern United States, the catchall phrases for any dogs taught to follow human trails were "blood-hounds" or "Negro dogs." They were instruments of terror, encouraged to be both trackers and attackers. The term "blood-hound" was an indiscriminate holding pen into which you could throw any number of dog breeds: hounds, foxhounds, bulldogs, mixed breeds. Harriet Beecher Stowe wrote about such tracking dogs in her classic *Uncle Tom's Cabin*. Not once did she use the word "hound" or "blood-hound."

Nonetheless, abolitionists were as aware as slave owners of the symbolic power of the dogs and of the term itself, which represented the horrors of slave tracking and slaveholding. The iconic illustrations in magazines, newspapers, and flyers of the era didn't resemble the bloodhound of today; instead, the depictions matched what was known as the Cuban bloodhound—a mastiff-like war breed, brindled, with clipped ears and broad heads and snouts. Cuban bloodhounds were a powerful symbol: These were the same huge dogs imported by British forces to Jamaica in 1795 to suppress a slave revolt. General Zachary Taylor, to his everlasting regret, approved importing them into Florida to track and attack Seminole Indians. The dogs weren't any good at finding the Seminoles, but they did create a public outcry, so they were removed. As historian John Campbell noted, southern slaveholders who used the Cuban bloodhounds provided abolitionists with great evidence to condemn slave chasing.

The ensuing decades have seen highs and lows for real bloodhounds in the United States. One of their principal working uses was for tracking escaped prisoners. Those dogs—tall and lean and less wrinkled than the AKC version—were usually trained and handled by prison trusties, a good number of them African-American. Many a trusty would be freed from prison, wrote bloodhound specialist Leon Whitney, and find a criminal excuse to make it back so he could continue working with his beloved bloodhounds.

The bloodhound's star rose with increasing use in police departments during the mid-twentieth century, and a few became celebrities. When Nick Carter, a tracking bloodhound, and his handler, V. G. Mullikin, would arrive on the scene in Kentucky, wrote Whitney, "great crowds gathered, so many people that they often constituted his chief problem—how to get started, rather than how to follow the trail." One of Mullikin's longest trails was reportedly fifty-five miles. He had to stop in the middle while one of his dogs had puppies. He sent her and the pups home and kept tracking with another dog. When James Earl Ray, the killer of Martin Luther King Jr., escaped from prison in 1977, bloodhounds tracked him for three miles, finding him in a pile of wet leaves.

The modern bloodhound, in other words, was evolving into a tracking machine that police and search-and-rescue teams still depend on today.

•   •   •

Every dog handler, law enforcement officer, and volunteer searcher seems to have a story about Andy Rebmann, who has trained dogs for more than forty years. Since retirement as a trainer for the Connecticut State Police, he's been teaching across the world, from Japan to Germany to Mexico. He's trained dogs and their handlers for trailing, patrol, narcotics, explosives, arson, and cadaver work. He's a court expert and an author. He continues to train bloodhound handlers. His own bloodhounds tracked hundreds of criminals and lost victims.

In 1972, Andy had been a state trooper for less than two years when he decided to try a patrol dog. A year later, he got a bloodhound, Tina, and fell in love with her nose and her trailing ability. Yet Andy, never sentimental, was noting an irritating tendency in his bloodhounds if they smelled a hiker who had gone beyond hypothermic: The dog would stop, looking hapless. Trail? What trail? Tina did that on her first dead person in 1973. "Not working into deceased subject," Andy's notes read. Even indomitable Clem—whose famous nose was upheld four times by the Connecticut Supreme Court; who trailed one man on an eight-day-old trail; who got a national award for his tracking nose; who was quite capable of tagging a felon with his teeth once he found him at the end of a trail—was a chickenshit when it came to dead bodies. He refused to trail all the way into them. The one time he did, he turned around and ran out the same way he had tracked in. "He almost turned me upside down," Andy said. "No way he was going to stay and sniff that guy."

On some cases, Andy had to tie his bloodhound to a tree and go poking around in the heavy brush himself. It was annoying.

4

Birth of the Body Dog

This animal exhibited a remarkable ability to detect all forms of buried explosives, and a surprising willingness to work with man. Were it not for the great size of this particular breed (400 pounds or more) and its unfortunate social habits, it might have been the ideal choice for detection service.

—Report #2217, U.S. Army, 1977

Founded in 1947 on the outskirts of San Antonio, the nonprofit Southwest Research Institute is dedicated to developing breakthrough scientific and engineering technologies and practical research that translate into immediate benefits for its funders, from oil and gas companies to NASA and the Department of Defense. The institute still designs spectrographs for missions to Mars, antidotes for chemical weapons, and compressors for offshore oil rigs. SwRI also plays around with the kind of wacky animal research that makes you think, with affection and wonder, "only in America."

This genre of blue-sky research wouldn't surprise anyone looking at the biography of the institute's founder, Tom Slick Jr., a Texas wildcatter, inventor, and committed cryptozoologist. He paid for three separate expeditions to Nepal to search for the Abominable Snowman of the Himalayas. He tried to get permission from the Nepalese government to use tracking bloodhounds, but the country refused to let the dogs in. Slick died in a private plane crash in 1962 at the age of forty-six, but his dream institute, SwRI, thrived, with brilliant scientists and engineers flocking to San Antonio. Today it has a staff of three thousand, one of the largest nonprofit applied-research institutes in the nation.

Slick's passion for mythic cryptids was surely an embarrassment for behavioral scientists at SwRI, who were deeply ensconced in training projects with real-world species. Yet something of his spirit lives on in their work. The institute's honeybee research in 2001, for instance, falls into a similar category. Even before 9/11, SwRI scientists were working on "a controlled biological system"—that's institute-speak for a cooperative critter—to detect bombs. Dogs aren't the only creatures who can be harnessed to help humans. The scientists trained the bees of twelve hives, giving them sugar-water rewards. The worker bees performed beautifully in field tests, buzzing right to their bomb targets and ignoring nearby flowers. Nor did their delicate bees' feet trigger explosions. SwRI researchers even put radio transmitters the size of a salt grain on the bees' backs, to track them as they honed in on distant TNT. The researchers thought they might be on to something big.

It was an inspiring experiment, but bees have their limits. They tend to die sooner than dogs, with a life span of about six weeks during high pollen season. They hate the cold, the dark, the rain. Using them at an airport security checkpoint isn't practical. I know this because David and I keep hives in our yard in North Carolina. Our bees hate three things we love: garlic, wine, and bananas. We can't consume any of those products before inspecting the hives, or we risk their displeasure. We love our bees, and need them for pollination, but I'd rather train Solo than a bee.

The idea that bees might have potential in both war and peace was the continuation of a long tradition, not just at SwRI but nationwide. The tumultuous early to mid-1970s were an enormously fertile time for detection research generally and research using animals for detection in particular. The Vietnam War was winding down. At the same time, Department of Defense–funded researchers noticed the skills of military dogs and wondered what else dogs might be capable of. There was enough intellectual and experimental curiosity—and money—to percolate from the military labs on the East Coast clear to San Antonio, Texas.

Nick Montanarelli, now retired but then a project manager at the U.S. Army Land Warfare Laboratory, remembers that era clearly: It was just a few years before he went on to co-develop the bulletproof Kevlar vest with Lester Shubin, an invention that continues to save thousands of lives. But at the time, Nick and a small cadre of other researchers across the country—including veterinarian and behavioral research scientist Edward E. Dean, behavioral psychologist Daniel S. Mitchell, and William H. Johnston with the U.S. Army Mobility Equipment Research and Development Center in Virginia—were starting to work together on detection projects.

That era, Nick said, was special: You could launch ideas, get results, and have an application in the field in six months. The military would provide up to three thousand dollars and tell Nick to start solving a problem. Often, he said, he'd fly out to San Antonio to brainstorm projects with Ed Dean. "I was down at Southwest Research every other week," Nick recalled. "Dean and I would go to lunch, and we would try to devise some methodology for trying some things."

SwRI and various army laboratories and centers worked together and separately, trying to determine how good dogs were at detecting land mines, punji pits, and trip wires, all of which were killing citizens and soldiers in Vietnam. There were new problems at home: assassinations of prominent political figures during the 1960s, from John F. Kennedy to Martin Luther King Jr., and during the 1970s, bombings protesting the war, as well as airplane hijackings. Could dogs be used to help find bombs in convention centers and guns at airports?

Jim Polonis, a project manager at SwRI for thirty years, helped manage a number of the successful and even some of the not-so-successful animal behavior projects. Like Nick, he has fond memories of those chaotic, fertile times. If someone had an idea, he said, talented researchers, trainers, and handlers were there to try to realize it. Jim Polonis's job was to make sure that when the ideas got turned into projects, everything went smoothly. It could be a challenge, hauling dogs from one end of the country to another. One spring, he, his wife, and their two children dodged killer tornadoes in a pickup while hauling a forty-foot-long horse trailer filled with German shepherds and Labradors from Fort Belvoir, Virginia, to San Antonio. Jim took care of dogs, handlers, and researchers on test sites across the country. One winter, he helped run mine-detection tests while fighting hip-deep snow and blizzards in Wisconsin. He and SwRI employees commandeered a utility truck with an attached telephone-pole digger to break test holes and plant mines in the frozen midwestern ground. Another year, he had to figure out how detection dogs and their handlers might cope with dust storms and 118-degree temperatures in Arizona.

Dogs weren't the only potential detection species to interest SwRI and the military. They added pigs to the mix, which wasn't actually much of a stretch: The Italians and French had used pigs to find pricey truffles since the fifteenth century. SwRI used red Durocs, an old handsome breed with drooping ears and the mahogany coloring, if not the fine feathers, of Irish setters. Unlike Irish setters, Durocs were exceptionally mellow. Jim Polonis remembered one that could detect buried mines at much deeper levels than any dog could. "That pig could detect anything," Polonis said. Partly, he thought, the pig wanted to please its talented, petite female handler. The final army report on mine detection, sadly, didn't give full credence to the gender of the best pig handler; instead, the report noted the pigs' great willingness "to work with man." No, that particular pig was clearly willing to work with woman.

There were just a few problems with the test Durocs: They were pigs, with "unfortunate social habits" and a certain stigma: "Would you let a German shepherd in to search your house or a red Duroc?" Jim asked me. Another problem was that red Durocs were highly regarded for their "excellent rate of gain," a plus for slaughter but a minus for mine detection—especially when the four-hundred-pound pigs got excited about finding mines. They would pull on their leashes. "They'd really drag you around," Jim said. More problematic than their enormous girth, and potentially more dangerous, was the pigs' "irrepressible desire to root in the soil," which needed to be discouraged during a mine search. So even though domestic pigs were especially effective at sniffing out all sorts of materials, SwRI ultimately rejected them as sniffer animals.

The test pigs weren't wasted. Joan Johnston, whose husband, William Johnston, was a researcher with the U.S. Army center that co-sponsored the study, remembered the great picnic SwRI hosted the year of the pig study: It featured a delicious pork barbecue.

The experimentation didn't stop with pigs. Coyotes, coyote-beagle crosses, deer, javelinas, raccoons, foxes, a badger, coatis, timber wolves, a civet cat. *Three* kinds of skunks: spotted, striped, and skunk-nosed. And the occasional indigo snake and rattlesnake, thought uniquely suited for mine detection because of an unsurpassed ability to crawl into holes. Researchers even tried raptors for mine detection.

The behavioral scientists, project managers, and trainers at SwRI were beginning to realize, with some disappointment, that wild animals had issues: They were wild. Wolves and foxes considered people "a menace to be avoided." The raccoons weren't awful when they were young, but as soon as they became adolescents, they started rebelling: They bit. The teen javelinas wouldn't listen to or perform for their human handlers. The coatis (a cousin of raccoons), despite their great snouts, were "lethargic." The deer couldn't search systematically. The rattlesnakes weren't prone to biting; they simply fell asleep in the sun.

The dog, ordinary *canis planus,* became the fallback. The dog might not have the most wonderful nose of all the animals in the kingdom, nor is it necessarily the most intelligent. It can't slither into tiny holes like a snake or leap over obstacles as nimbly as a deer, but it can go a lot of places. It's the right size for a number of tasks. It can walk at your side. The dog lives long enough to make the training worthwhile. It isn't nocturnal or diurnal but is happy to be awake when you are. Above all, the dog wants to please.

The dog, SwRI and the army concluded, was just right.

"We found the dogs so useful," Nick Montanarelli said. "That's how I got into detection work."

That wasn't the end of tweaking and experimenting. Both SwRI and the army tried a number of dog breeds for a multitude of purposes and climates, from Australian dingoes and Norwegian elkhounds to Dobermans and cocker spaniels. The scientists and project managers didn't experiment with mutts. The reason had nothing to do with blue-blood snobbery or the German propensity for creating an über-shepherd that could best represent the nation-state. It was a science thing. Mutts couldn't be replicated easily, and if you were trying to get uniform results, you needed consistency. *Canis planus* isn't that plain at all, genetically speaking. While other breeds worked fine, the best all-around dogs for a multitude of tasks were the German shepherds and Labradors. They had hunt drive and play drive. They were the right size. They had fine noses. No one would laugh derisively at the handlers. The two breeds weren't without faults, especially the German shepherd. Even then, the army was trying to develop a solid breeding program for shepherds, to make them confident and capable of being aggressive when needed but not nervous and without hip problems—sadly, both propensities that American-bred shepherds, thanks to Rin Tin Tin's popularity, were starting to exhibit.

There were exceptions to the shepherd-Labrador dominance for particular project needs. In 1971, when D. B. Cooper hijacked a plane and parachuted down over Washington, disappearing with two hundred thousand dollars in ransom money and inspiring a rash of copycat hijackings, SwRI and the military labs came in with a solution. A soigné woman with a lapdog draped over her arm would walk through an airport terminal and boarding areas. She would pass close to the waiting passengers. If the dog smelled a handgun, it would scratch at the woman's arm. Polonis recalled trying Lhasa Apsos, miniature greyhounds, and whippets, among other small breeds. The whippets, he said, outsniffed them all.

Some of the other small breeds didn't work out as well. Happily, not all of the institute's failed experiments ended with a barbecue. William Johnston brought one of the lapdog dropouts, a Maltese puppy too pug-nosed to sniff properly, home to his wife, Joan, and their children in Virginia. Puffin lived with the family until she died at the ripe old age of thirteen.

. . .

The next time you visit a zoo or a natural history museum and survey the extraordinary diversity of the organisms on our planet, pause for a second to remind yourself that all this variation—the elephant tusks and peacock tails and human neocortices—was made possible, in part, by error.
—Steven Johnson, *Where Good Ideas Come From,* 2010

Good ideas can occur separately and seemingly in isolation, like the summer popcorn storms in North Carolina that arrive in late afternoon to water the vegetable garden. What good ideas and rainstorms both need are the right atmosphere and some basic ingredients. So it was that around the same time, several researchers, trainers, and agencies appeared to independently arrive at the idea of the "body dog": using dogs to find victims of homicide, disaster, accident, or war. That idea was a natural outgrowth from other ideas but also from errors and dead ends, such as the pig experiments.

One of the earliest efforts came in July 1970, when the Lancashire Constabulary in England started training dogs to recover the dead. Its training program, using pig meat as a substitute for human tissue, lasted eighteen months. It worked, according to the few accounts available, although the history of the program is spotty. An English handler and his dog were deployed in the Sinai after the disastrous Yom Kippur War in 1973. Tasked with recovering the Israeli dead, that team reportedly recovered 147 bodies, and an Israeli dog-and-handler team hurriedly trained for the task found the body of Anwar Sadat's brother, a pilot.

Within two years of that project, Nick Montanarelli had launched the first U.S. body-recovery dog study out of the U.S. Army Land Warfare Laboratory. The study wasn't an idle, let's-see-what-we-can-dream-up experiment. Nick was a practical guy, already thinking beyond the Vietnam War. Dogs, he thought, might have been useful in two domestic disasters: Hurricane Camille in 1969 had killed 259 people as it ripped through Cuba, up the Mississippi Delta, and flooded Virginia. The hurricane's actual wind speed was never known, as it destroyed all the recording instruments. In 1972, a dam in South Dakota's Black Hills broke, sending water cascading down a creek into Rapid City. Two hundred thirty-seven people died in a matter of hours, many buried under mud or swept away. Nick also had been talking to handlers and trainers in Canada, where the search-and-rescue dogs weren't doing a good job of recovering the dead in avalanches.

Finding appropriate training material for body-recovery dogs was a challenge. Nick's position in the military, with its long history of honoring deceased servicemen, kept him from using human tissue for training. Nonetheless, he wanted to get as close as he could to the real deal. He visited morgues and talked to morticians. He talked with military people who had been around lots of bodies. His solution was a combination of sweaty soldier uniforms and monkey meat (or as the report called it, "macerated subhuman") with some other chemicals added. It was a potent mixture, Nick recalled. The dogs found it. The four German shepherds in the study learned to work in fields, in buildings, in rubble, with a 92 percent accuracy rate in the final tests. Nick signed off on the study in May 1973, and the dogs went on standby for disasters.

It's difficult to trace the exact relationship of who did what and when because of gaps in the record. People die. Memories fade. Some of the work was classified. But at the time Nick finished his report, Southwest Research Institute, too, was studying whether dogs could help find the dead.

The record of what happened next is clear. To make the leap from speculative military research in Texas and Maryland to paws-on-the-ground cadaver-dog work took not a major hurricane or flood, as Nick might have predicted, but a brutal murder in New York's southern Adirondacks. Mary Rose Turner, a mother of five suffering from depression and insomnia, left her house in the wee hours of April 26, 1973. Her walk led her past Bohling's Shell station in rural Syracuse, where a man named Bernard Hatch was working the graveyard shift.

Later that morning, a witness saw a car dragging what he thought was a "six-foot-long white object." It bothered him, so he didn't let it rest. He brought the New York State Police to Potato Hill Road in Steuben, New York, to investigate. The tissue and blood trail was more than nine miles long. Police found the rest of Mary Turner three days later in a shallow grave. Her body had been mutilated not only by the dragging but by dismemberment.

Evidence slowly and inexorably piled up against Bernie Hatch. A grand jury indicted him of Turner's murder on October 17, 1973. That wasn't the end of it. Just a month and a half after his indictment, hunters found the skeletal remains of Linda Cady, twenty-two, and her daughter, Lisa Ann, three, in shallow graves. It was two and a half years after their disappearance. They were just a few hundred yards from Mary Turner's grave.

The relationship among the victims, the location, and Hatch appeared more than coincidental. Cady and Hatch had dated for many months, with Cady joyfully noting in her diary that Hatch had given her a diamond ring. Authorities begin to suspect the area off Potato Hill Road was a burial ground, and they realized that Hatch was connected not only to Cady and her daughter but to another missing woman and her children. In mid-December, as searchers scoured the area, they found children's charred clothing not far from Turner's grave. The family identified the clothes as belonging to the three young sons of Lorraine Zinicola, who had also dated Hatch. She and her sons had been missing since September 1971.

The New York State Police put out the call, and on December 21, 1973, William H. Johnston flew in from the Military Animal Science program at Southwestern Research Institute in San Antonio to the little town of Steuben. He looked at the terrain and search conditions with the state police. Could the military dogs they were training to find buried bodies be used to find other possible victims of the now-indicted Bernie Hatch?

Investigators turned to a handler living 125 miles away: the New York State Trooper Ralph D. Suffolk Jr., aka Jim, who had a stellar reputation as a bloodhound handler. He and one of his dogs, Colonel of Redstone, were already renowned from having run a long trail that helped police locate three robbers a few years before—a conviction that was upheld in New York criminal court in 1969. This had been a first for tracking and trailing dogs in New York. The only legal precedent in New York for using a canine to help convict someone had not ended so well. In 1917, the New York State Supreme Court overturned the sentence of a woman convicted of arson based on a German shepherd's nose. That dog, the court declared, simply showed off for houseguests.

Jim Suffolk's bloodhounds didn't show off for guests. They tracked people almost daily. To Suffolk's great credit, he admitted under oath that bloodhounds were not infallible. His honesty garnered him more credibility.

The scientists at SwRI had wanted to use dogs to work domestic crime scenes, and Suffolk was the ideal man to take the lead. For this job, his trusty bloodhounds wouldn't do the trick. Though they were ideal for tracking the living, he needed dogs trained to locate the dead.

Suffolk flew to San Antonio in early May 1974 to start working with a newly invented category of search dog: the body-recovery dog, or body dog for short. Suffolk made training and handling suggestions; SwRI researchers made other recommendations; and Suffolk returned to upstate New York with two dogs. Pearl, a sweet-looking, white-blond Lab, was the furthest thing from a classified military project one might imagine. She was five years old in 1974 when she landed in Oneida County to take up the hunt for a possible serial killer's victims. Every snapshot of her has her mouth open, gazing lovingly at either Jim or whoever was behind the camera. Maybe that someone might slip her a cookie? Pearl had been shipped from place to place to train for narcotics, bombs, and land mines. Her last trained specialty was buried bodies.

Her sidekick, Baron Von Ricktagfan, a muscular black-and-tan shepherd, had already trained as a military scout dog in Fort Benning. Baron had a snarky edge to him, Jim Suffolk recalled. He could also do the work.

Jim Suffolk and his two new body dogs hit the woods of Oneida County, New York, starting in the patch of land where several of the victims had been found. They searched for additional shallow graves for the next seven months, until snow and ice in November stopped them. The search included four thousand acres of land, most of it pinewoods planted in the 1930s. They spent eighty-three days in the woods.

The search was interrupted occasionally for more pressing police business, including Pearl's search for bombs in the Oneida County airport before Vice President Gerald Ford landed there. Pearl found nothing except the training material planted by the Secret Service to make sure she knew her bomb business. Jim and his dogs also went out to a sewage treatment plant in nearby Onondaga County after a sewage worker admitted to raping and burying a Syracuse University student there two years earlier. Both dogs alerted on the same spot. The Syracuse police brought in a bulldozer and found Karen Levy buried a few feet down. Afterward, Suffolk studied his records and the terrain. The dogs were about fifteen feet off, on the downhill slope, probably because of an underground creek. He thought he should have insisted on continuing to work up the hill even after the dogs alerted. This is the kind of knowledge that would help forensic anthropologists understand the patterns of cadaver-dog alerts around clandestine burials.

Bernie Hatch, despite being suspected of killing a total of seven women and children, was convicted only of Turner's murder after a seventy-day trial, the most expensive in the history of Oneida County. He is still in prison in Auburn and says he is innocent. Jim and his two body dogs, despite months of careful effort, never did locate additional buried victims. Lorraine Zinicola and her three young sons were never found.

While Jim Suffolk and his dogs didn't find more bodies in connection with the Hatch case, his contribution to the history of working dogs was significant: the first fully recorded occasion when body dogs were used in the United States.

That was only the beginning for Jim Suffolk, who worked with Pearl and Baron for years. He used "the real stuff" to keep their training up: Bodies lying out in the woods tend to produce good training material. Jim retired from the state police in 1986 and is now in his early eighties. Despite recently losing a leg to circulation problems, and repeated shoulder surgeries as a result of holding on to harnessed bloodhounds running up and down hills in the Adirondacks, he is a local town justice. He and his wife, Sally, live in a house overlooking Canadarago Lake.

I couldn't resist asking Suffolk the obvious question: Did he consider using bloodhounds? I knew he would admit that shepherds and Labradors were better suited for cadaver work.

"Use the bloodhounds for cadaver?" he responded with horror. "Heck no! That was a colossal waste of a nose." On the other hand, he admitted upon reflection, while he believes bloodhounds have the Cadillac of noses, they aren't great at hopping around and getting into tight corners. Then his voice got wistful. "I've always wondered if I could train a cadaver cat."

The army and SwRI researchers, ever optimistic and open-minded, had already been there and tried that. Cats didn't care to communicate with the researchers about whether bombs were close by. "Cats were excluded from the final programs because of their demonstrated refusal to cooperate consistently in joint ventures with man."

. . .

I wasn't qualified until I said the two magic words, "Andy Rebmann," and they said, "My God, you trained with Andy Rebmann? You can search."
—Edward David, deputy chief medical examiner, Maine, 2011

It would be inaccurate to say that the cadaver-dog world was empty and void before Andy Rebmann arrived on the scene. Other people came before him: Nick Montanarelli, the researchers at SwRI, and, of course, Jim Suffolk. That same decade, William D. Tolhurst, a famous bloodhound trainer and handler in New York, noted in his memoir that he started training his bloodhound, Tona, in 1977 as both a "man-trailer" and a body dog. Other handlers and trainers are probably lost in the spotty records.

The world of the body dog was, though, still somewhat formless.

Andy was at a cop conference in the mid-1970s when Jim Suffolk gave a presentation on body dogs. Andy, fascinated, approached Jim afterward. The two men must have made a distinctly odd couple. Jim Suffolk looked like a burly James Garner, with an even more heroic chin. He filled out his immaculate state trooper's khakis. He had a full head of dark wavy hair, usually confined under his trooper's hat. Because he was at a conference, Andy might have been in uniform and not wearing his trademark faded baseball cap pushed up at an angle, exposing his large eyes, mobile mouth, slabs of facial plane, and magnificent ears. He probably was smoking a Pall Mall.

Andy introduced himself and asked Jim how he trained the dogs. Jim refused to tell him. It was still classified information. Andy didn't take the refusal personally; he knew Jim was working with a military research group. It simply stoked his determination. "Goddammit," Andy said, "I was going to have a body dog."

So Andy went his own way. He talked to a pathologist at the Connecticut Department of Health, who suggested that he start with the noxious chemical compounds of cadaverine and putrescine. Redolently available when animal tissue decays, they had been identified and isolated for over a century, having first been described by German physician Ludwig Brieger in 1885. Neither is the exact equivalent of human decay, since some stinky cheeses and even bad breath contain those compounds, but these were the early days of cadaver-dog training and human decomposition science.

In the mid-1970s, just as Andy was working out the scent of human death with the Connecticut pathologists, a cognitive psychologist at Tel Aviv University, Robert E. Lubow, was honing in on a question about the Lancashire Constabulary's program and the U.S. military program: "We must return again to the problem of stimulus generalization," Lubow wrote in his fascinating book *The War Animals*, published in 1977. "The British trained a pig detector, the Americans a monkey detector. What evidence is there that these dogs, each trained to a very specific odor, will be able to generalize to the real world human body odor task?"

That basic question would plague all sniffer-dog work—not just cadaver detection—for decades to come. Early in his training, Andy had speculated that decomposing animals and people smelled pretty much the same. Soon he, like Jim Suffolk before him, realized there were significant differences. Training dogs on the real thing was ideal, and law enforcement didn't have quite the same constraints about using human remains as the military did. After crime or suicide scenes were processed, something always remained that could be harvested to help train the dogs.

Rufus, Andy's stocky, dark German shepherd patrol dog, had started out as a potential guide dog for the blind at the Fidelco Guide Dog Foundation in Bloomfield, Connecticut. Rufus had flunked out of the program because he wasn't suited to calmly and gently guide anyone. He was a fine patrol dog, but Andy also started training him on a combination of cadaverine and putrescine but also with the dirt harvested underneath bodies that contained fluids and adipocere (the waxy fat that persists in some environments). That was in 1977, the same year Andy's other dog, Clem, a bloodhound, won the award for being the best man-trailer in the nation. As long as people were alive, Clem was happy to find them. If they were assumed to be dead? Rufus took over. It was a sweet deal for both dogs.

By 1980, Andy was starting to use Rufus and his nose for increasingly challenging work. Clandestine burials are the worst cases to work. They involve shovels and bulldozers and sometimes jackhammers. If you are off by more than thirty feet, a body might as well be a mile away. No one likes to dig to find bodies. Especially without a lot of corroborating evidence.

Everything looked perfect and orderly at the neat, raised-level, shingle-sided ranch house in Monroe, Connecticut. It had a manicured lawn, a round swimming pool in the backyard, and a new patio next to the pool. Robin Oppel, twenty-eight, had disappeared. Her husband, Kent Oppel, a twenty-nine-year-old self-employed businessman, had given the police permission to search the premises without a warrant.

Robin's car had been found abandoned twenty-five miles away nearly a month after she disappeared. Inside the car was a broken-off portion of an owl key ring but no sign of Robin. She was seen last on September 19, 1980. At first Andy thought he might start one of his bloodhounds to see if the dog, even after all that time, could pick up a trail and give them a direction of travel. That was a stretch, and he knew it. By then, detectives working the case had a hypothesis.

Rufus had been working as a body dog for three years when he and Andy arrived at the Monroe house to search. While Kent Oppel watched, Andy started Rufus on the front lawn, then down the side of the house and to the rear of the lawn. Rufus walked along the fence toward the swimming pool, stuck his nose in the dirt next to the newly laid concrete patio, and started digging. That was it. Andy walked him away and shrugged casually. He could hear Oppel telling bystanders that the dog obviously hadn't found a thing.

For long terrifying minutes, Andy thought perhaps Rufus had screwed up. Investigators jackhammered the concrete next to where Rufus had indicated, dug down a foot, and ran into electrical wires. Andy brought the dog's nose back in. Rufus, Andy recalled, started "digging to China." Investigators kept shoveling. Just a little farther down, they saw a small object in the hole: the other half of the plastic owl key ring found in Robin's abandoned car. They kept going. Robin was four and a half feet down, under the concrete, beneath a layer of lime powder.

Because the body dog represented a new and fascinating canine career, sometimes journalists got the terminology wrong. Rufus, one newspaper reporter noted with great sincerity and inaccuracy, was "one of eight 'dead dogs' in the country; the only one in New England." Such reports of Rufus's death were premature. He recovered twenty-six bodies in his career.

Andy Rebmann and Jim Suffolk's relationship didn't end at that cop conference. A picture of the two men in a 1986 Vermont newspaper shows them using their body dogs on a homicide search. Jim looks spit-shined in his pressed khakis, facing straight into the camera with a nice smile. Andy looks informal if not disreputable, in jeans, T-shirt, baseball cap, grinning broadly, sideways to the camera. Jim Suffolk's shepherd was a big male called Argus. Rufus's successor sat next to Andy: She was a delicate, light-boned German shepherd he had named Dupa—Polish for "ass" or "hot chick." After a missing persons search in a Polish neighborhood, the Connecticut State Police made Andy rename her, and after that, he called her Lady.

Like Rufus, Lady ended up earning her kibble and taught Andy more about dogs' ability to find the dead. In mid-January 1987, it wasn't a buried body but a body spread far and wide—a case that involved forensic scientist Henry Lee (who later became famous as a defense expert in O. J. Simpson's trial). The case became the inspiration for the darkly comic Coen brothers' movie *Fargo*. Helle Crafts, a flight attendant, was missing after beginning divorce proceedings against her philandering husband, Richard Crafts, an airline pilot. Crafts had used his credit card to rent a wood chipper and to buy a freezer and a chain saw. A snowplow driver reported seeing a man using a wood chipper along the bank of Connecticut's Housatonic River in the middle of the night during a snowstorm.

Lady was put to the tedious task of sniffing piles and piles of frozen wood chips hauled in from the riverbank. One pile was particularly interesting: Lady alerted. It's here. What she had found, although tiny, was human. Ultimately, because of Lady's alert, police recovered sixty tiny chips of bone. A bit of blood. Strands of blond hair. A tooth with a gold crown. And a fingernail whose color exactly matched a bottle of polish in Helle Crafts's bathroom cabinet. It was the first time in Connecticut history that a murder conviction was secured without a body. Richard Crafts was sentenced to fifty years in prison in 1990. The earliest he can be released is August 2021, when he'll be eighty-four years old.

. . .

Another object [of this invention] is to provide a method of building construction in which the dangers attendant upon working at elevated levels will be reduced to a minimum.

—United States Patent Office No. 2,715,013, August 9, 1955

It was just three months after the Helle Crafts case, in late April 1987, when Andy used cadaver dogs for the purpose that the U.S. Army and the Southwest Research Institute originally envisioned: a major disaster, the worst in modern Connecticut history.

Twin sixteen-floor concrete buildings under construction in Bridgeport collapsed. L'Ambiance Plaza fell in seconds. Within hours, Andy and Lady, along with four other Connecticut state troopers and their dogs, were on the scene, facing a mountain of broken concrete slabs, twisted steel, and iron rebar. Dogs and men inched across the pitched slabs of concrete. The troopers and workers carried spray paint and flags. The German shepherds would alert, giving the general location of body after body, sometimes on open holes, sometimes at the edge of the pancaked slabs, where scent could escape.

Though it was early on, construction workers and their families were becoming aware that the scene was less a rescue operation than a recovery operation. Twenty-two workers were injured, some badly, but they were the fortunate ones: blown off the edges of the slabs by the force of the collapse as the floors pancaked down. The dogs alerted time after time after time, inhaling concrete dust. Then the cold spring rain started, tamping down the dust and making footing even more treacherous, intensifying the cold glare of the floodlights on the massive rubble pile.

The dogs helped find all twenty-eight victims. Italian-American, African-American, Irish-American workers, their bodies so broken that Andy said he had never seen so much damage on human bodies, before or since. And Andy has seen almost everything that humans, or nature, can do. "It still haunts me," he said.

L'Ambiance Plaza still angers him. Quick, cheap—and dangerous. It still angers me. In a minor twist of fate, Andy and I realized nearly a quarter century later, when we met face-to-face, that we had probably passed each other on that site. Andy was managing the dog searches for days, until the last body was removed. I was there only one night as a newspaper reporter for the *Hartford Courant*. All disasters, by nature, are terrible, but it was the worst disaster I had ever covered. I played only a brief role one freezing day and night, tasked with standing by in case more injured victims, or bodies, were recovered. Reporters and investigators came to understand that something had gone disastrously wrong with a construction system hailed as an efficient, economical way to raise a building. After the accident, lift-slab construction was temporarily banned. The ban is no longer in place; nonetheless, lift-slab construction is rarely if ever used in the United States.

That Andy and I didn't meet during those terrible days in Bridgeport probably didn't change the course of my life. I doubt I would have decided, at that point in my newspaper career, to start training dogs to do search work. That would wait until I was solidly middle-aged.

The odd connections didn't end there. As I was finalizing my research on SwRI and its role in dog research, I discovered an unrelated invention of its founder. Tom Slick Jr. filed a patent for "Apparatus for Erecting a Building" in 1948. As the old pen-and-ink drawings filled my computer screen, I saw the original outline of a construction system I had grown to know by heart in the aftermath of L'Ambiance Plaza: its pulleys and jacks and pumps and concrete slabs. Slick had invented lift-slab construction.

Andy said it best: "What a coincidence! He funded the type of dogs that would work to locate victims of a disaster caused by the failure of his invention."

Andy, to no working-dog person's surprise, ended up writing (along with Edward David and forensic anthropologist Marcella Sorg) what is considered the bible of cadaver-dog trainers and handlers: *Cadaver Dog Handbook,* published in 2000.

"Remember I warned you about being too brain-oriented. The Andy Rebmann book is good, though—he's the guru," Nancy Hook told me in an e-mail.

Andy's wife, Marcia Koenig, a famous volunteer handler and trainer in her own right, helped write and edit and provide illustrations for *Cadaver Dog Handbook.* She had been doing search-dog work since 1972. Andy introduced her to cadaver-dog work. Marcia became very good at it. She and her German shepherds have deployed to look for missing homicide victims, suicides, lost hikers, dementia patients, and victims of tornadoes and hurricanes. She's worked in wilderness, in snow, and on water. She and her sable German shepherd, Coyote, spent four days in August 1997 crawling through mud on the island of Guam after Korean Air Flight 801 crashed and tore a ragged hole in the side of the mountain.

"That area was so saturated with the smell of decomposition and jet fuel that none of the dogs could alert on anything specific," Marcia recalled. "Each one looked up at the handler in frustration and basically said, 'It's everywhere.'" She and Coyote, as well as the other dog teams, were knee-deep in wet clay during the entire search. Despite the challenges, stubborn Coyote helped find bone, tissue, and a femur. Toward the end, when the mud was too deep and Marcia exhausted, Coyote, a wild and crazy dog gone good, laid an object at her feet.

"She was so gentle," Marcia said. It was a child's foot, nearly the last thing found on the search. Retrieving wasn't standard operating procedure for Coyote, but that little foot gave the searchers and the family great comfort.

Across the country dozens of handlers and trainers have trained with Andy, then followed in his footsteps, training their own dogs, and also training other handlers. In law enforcement in the United States, Jim Suffolk started the legacy of body-recovery dogs. Andy kept it going and developed the training system considered the gold standard today.

Andy is in his seventies now, still traveling worldwide with Marcia: to train dogs and handlers, to create better training protocols, to testify in legal cases. He still goes out on both live searches and cadaver searches. He can't tell me how many searches he has done over his career. I know that before his retirement from the Connecticut State Police in 1990, he used to do at least a hundred searches a year. He doesn't keep count anymore. What would be the point?

"My next search is the important one."

5

The Shell Game

... the foundation ceremony can be seen as a recognition that building is
both an act of memory and also a fresh start.

—Tracy Kidder, *House,* 1999

Solo stood in Nancy Hook's training yard, brow wrinkled, staring
at five identical white buckets lined up with military precision. One
of them had a cadaver "hide" in it—a little bit of something from
somebody who died, or from someone kind enough to donate part of
himself to Nancy for a moment like this. One canine trainer shared a
portion of his rib, removed in a surgery. As he explained to me years
later, he didn't want his own dogs wandering around confused about

what they should be indicating on—the rib outside the handler or the ribs inside him.

For this particular exercise, Nancy was using one of her soon-to-be-ex-husband's wisdom teeth, packed with a bit of bloody gauze. Now nearly six months old, Solo was fifty-five pounds of raw bone and sinew. Few juvenile German shepherds are handsome. Only a handful are well adjusted. He was increasingly impervious to pain, whether receiving it or causing it. Despite my nightmares, Solo wasn't that far outside the shepherd mainstream.

Nancy had given him his first job: to duck his head inside bucket after bucket and figure out which one held the tooth and gauze. This is one method for laying a foundation for a cadaver dog, or any scent-working dog, as it learns to recognize and then signal clearly that it has found what you want it to find. Cadaver, cocaine, gunpowder, heroin. Bed bugs. Some trainers use buckets; others use concrete blocks. More advanced rigs consist of wooden boxes with holes in the top, even springs inside so that a rubber Kong or tennis ball can pop out like a jack-in-the-box for an instant reward. These were early days for me, before I knew the great varieties of boxes available. Everyone has a favorite system, but bells and whistles aren't necessary; perfect timing on the part of the handler is.

On the first run, Solo ducked his head into the fourth bucket, which held the bloody tooth, looked up at me, then ducked his head back in. He had no association with that odor, although it smelled intriguingly different than the smells in the first three buckets. Nancy hissed my cue at me, and I fumbled to give him a liver treat. Solo tried to help himself to the tip of my finger along with the treat. Soon enough, he threw himself into the game. As Nancy switched the position of the buckets, he charged from one to the next, jerking me along, tangling us in his lead, pulling his head out of the "hot" bucket, staring at me, griping loudly if I didn't reward him quickly. His complaints moved up and down the scales, howls of frustration and delight.

Nancy's chestnut eyes narrowed as she watched Solo and me perform a bad rendition of the funky chicken with some leash bondage added. I could hear my heart forcing blood through my head. It should have been simple. I was to move just ahead of Solo, using a loose lead, past each bucket, not hesitating, not rushing. With a gracious hand gesture, I was to present the bucket to him. Check here (dog's head dips into the bucket), check here (dog's head dips in the next bucket), check here (dog's head dips and stays). Good dog! Treat! Classic operant conditioning. Solo would start linking cadaver smell with a reward.

Nancy let me keep the treats in my handy belly pack, but it was turning out to be one more thing to manage besides the lead, the dog, the buckets. Oh, and my ego. I was terrible at this. Solo surged from one bucket to the next, skipping one that didn't seem interesting, doubling back, yanking us silly, then yowling when he got a whiff of scent and changed his mind. He was cheating, energetic, and out of control. Nancy loved it. She chuckled and crooned, "Good boy, good boy" to him while she hissed sotto voce to me, "Reward him, reward him."

I was near tears. I didn't fully understand then, but Solo was in what working-dog trainers call "drive" mode, as essential to keeping a dog running as gas in a car's tank. That revved-up state of mind would be essential to the work ahead. I saw it as "bad dog" mode; it wasn't what I was used to. Zev had walked quietly and steadily at my side in a perfect heel. He got depressed and shut down if I scowled at him. Even Megan, though she cared not a whit whether I approved of what she did, was obedience-trained. Their good behavior was a reflection on me. I had tenure. I was a teacher of dogs and humans.

Solo was brutally rebooting my canine worldview. According to Nancy, I had my first working dog. As far as she was concerned, any male dog who was interesting or worthwhile was a "macho jackass." Any good female was a "bitch from hell." These were compliments. Sweet, compliant dogs were boring, and Nancy wanted nothing to do with them. Solo was making me miserable while achieving comparative perfection in Nancy's eyes. I could feel her already skeptical assessment of my potential nosedive. I was not a bitch from hell. I was trying my best to be compliant, but I couldn't even coordinate my limbs.

"That's it," Nancy instructed when she heard Solo yodeling at me for about the third time because he wasn't getting rewarded. "There's his alert!"

If I hadn't been so frustrated, I would have marked this as a special moment in a working dog's life, like Solo getting his tribal name—"Whines with Brio." The behavior of an alert, or what some in the sniffer-dog business call a "final indication," is supposed to be something that comes naturally to the dog yet is distinctive. For most narcotics dogs, the alert means sitting with a focused stare at the spot where the drug smell is coming from. A few drug dogs still dig and scratch, though that "aggressive alert" is disappearing, seen as old-school. Bomb dogs never used it, for obvious reasons. Nancy and I talked: Solo's distinctive whine combined with a sit might become his trained alert to tell me he had not only found the material we were looking for, he was committing himself to that spot. More important, I could say on a search, "This is what Solo does when he detects the odor of human remains. He sits and sings a cappella."

That moment wasn't in our immediate future. If it ever came. This laying of the scent foundation and formulating an alert were the first baby steps on the long road to mastery.

It wasn't just the scent that Solo needed to recognize. He needed to be willing to go anywhere to find it. That meant turning his natural drive into environmental toughness. I was slowly starting to understand why working-dog trainers liked dogs who were pains in the rear, who destroyed crates, who tore up the insides of cars, who challenged everything, who tried to jam three toys in their mouths at once. The first time I met West Virginia working-dog breeder and trainer Kathy Holbert, she was in the yard offhandedly throwing a dog's rubber Kong. It landed repeatedly in the middle of a heavy brush pile that looked like a funeral pyre, the kind of awkward toss that makes most people curse. Only Kathy was doing it on purpose, and the young shepherd was diving into the rough limbs, making them part like water. Kathy was developing the dog's nose and drive.

The energy had to be there first. If you had that in a dog, you could work with it. As canine trainer Lisa Lit explained to a group of search-and-rescue handlers about building drive, "Let them explode, then rechannel it."

First comes the energy, then the expertise. Cognitive scientists have intensively studied the notion of human expertise. We watch playful children start out banging incoherently on the piano. That's a start, but it's the structured, guided practice and play with constant feedback over an extended period of time that can turn random notes on a keyboard first into "Doe, a deer, a female deer" and ultimately into Thelonious Monk's " 'Round Midnight." That is, if a parental figure doesn't ruin the sound of music by haranguing the child to practice. Along the way, a number of the motor behaviors for playing the piano become automatic, so the child doesn't have to think about them. The fingers start to fly by themselves up and down the ivories as body memory pulls them along.

The notion of expertise applied to dogs and other animals is scientifically controversial. Working-dog trainers have no doubt that it exists, and they aren't worried about whether the learning curves mimic humans', as long as the dogs learn, keep learning, and layer that knowledge.

The beginning of the process was what Nancy was teaching me and Solo: "What the heck is that smell?" That early stage is important. Once that scent is second nature, add some distractions: some of Nancy's chickens, perhaps. A clumsy handler like me. I might teach a juvenile Solo to teeter on a low balance beam in the backyard and to keep his paws on the board. I wouldn't comfort him if he fell off and yelped; I would urge him back in a happy, relaxed voice until he could walk along the beam with a sure paw and confident grin.

William "Deak" Helton calls the entire emerging research arena of working dogs "canine ergonomics"—the study of the relationship between the working dog and her environment. The working dog learns the gymnastics of the body, mind—and nose. In due time, a good disaster dog should be able to crawl and balance herself purposefully over the rubble of a collapsed building, all while using her nose, then signaling to the handler what she has found. That dexterity, that multitasking, is what Deak Helton calls "canine expertise." And Deak is a believer in the concept. "Although canine experts cannot verbalize their knowledge, this in no way implies they do not have it," he wrote.

There's resistance to the notion that dogs can be experts, among both cognitive psychologists and people who feel it gives any animal too much credit. "I think a major problem is the bogeyman of anthropomorphism," Deak said. Yet we can train people to do things that other people can't do without training. "If you asked me to do a backflip and discovered I could not do it now—untrained, unconditioned—would you conclude people cannot backflip?" Watch gold-medal gymnast Gabby Douglas, the "Flying Squirrel," at the 2012 Olympics and try to imagine her first gymnastics lesson at the age of six.

At the same time Nancy was teaching me how to train Solo's nose, I needed to teach him skills that would complement that nose: how to tolerate electric fences, how to swim in rivers, how to push through heavy brush, how to climb over and into and under and through. How to ignore distractions, like Whiskey, Nancy's big brindle East European shepherd—Solo's nemesis—who snarled next to the cyclone fence, suggesting that the upstart come try him out.

When it comes to dogs doing detection work, Deak said, people tend to forget it takes time to develop skills. Dogs, like humans, need a chance to learn before their capabilities are dismissed. I was going to give Solo a chance, even if he wasn't giving my fingers much of one.

Science aside, experienced working-dog trainers are clear about the order of things. You lay the foundation first. The analogy with building a house is perfectly apt. If you don't get that right, nothing you put on top will hold. You'll have a shaky, unreliable dog. A dog who can't keep his eyes on the prize.

. . .

> Alice started to her feet, for it flashed across her mind that she had never before seen a rabbit with either a waistcoat-pocket, or a watch to take out of it, and burning with curiosity, she ran across the field after it, and fortunately was just in time to see it pop down a large rabbit-hole under the hedge.
>
> —Lewis Carroll, *Alice's Adventures in Wonderland,* 1865

For the moment, Solo was staring fixedly at the buckets, ignoring me. Whatever fun thing was happening, it was coming from those buckets. Solo loved the buckets. That could change. While this kind of foundation work is essential to future stability, it can be boring for both you and the dog. Especially if you forget that it's supposed to be fun, which is easy for an overly invested handler (like me) to do. That's when a little magic can come in handy. One day, without Solo being along for the ride, I got to participate in a magic show and learn from a master trainer how to keep foundation work interesting for both handlers and dogs.

The shell game is probably one of the oldest cons in a world filled with bait-and-switches. Illustrations and detailed accounts of shell games date back to Greece, long before a deck of playing cards was the easiest way to mount a quick three-card monte game on a street corner. It's a game that includes misdirection and dexterity, audience participation and manipulation. On busy streets all over the world, hustlers are setting up shell games, using shills in the crowd to pull in gullible marks, who can't help watching and then can't help thinking they can beat the system. Soon the marks' money has disappeared, along with the hustlers and their shills.

Canadian Kevin George is a conjurer, a dog trainer, a people trainer. Before he was a master dog trainer, before he tried his hand training an elephant and a bear, before he was a cop, before he was a rodeo clown or a shiatsu massage expert, Kevin was a kid who loved magic, who loved learning how to make his fingers furl and then unfurl like birds set free, who plucked coins from behind ears. That love of magic stuck around, through decades of training dogs to bite bad guys, to search for drugs, to track lost children and criminals, to find them alive and sometimes dead.

Every solution in foundation training stems from a unique problem. Back in 1978, Kevin had a patrol dog who didn't want to search properly. It was all a big yawn to the German shepherd. Kevin wondered whether he could use magic to motivate his dog. He wanted to teach the dog to search with enthusiasm, to quarter back and forth properly, to cover the search area thoroughly. To be interested instead of cynical. To stick his head in confined spaces. It was a tall order. Kevin filled that order with magic cardboard boxes.

Since I was Solo-less at a dog seminar in Seattle, I was tapped to be Kevin's magician's assistant for his show, "Box Magic." Kevin practices the purest kind of magic: the art of misdirection and the craft of sleight of hand. He teaches people to teach their own dogs. A trainer's job is to try to train the handler, and the handler's job to motivate the dog. But the handler has to be motivated in order to motivate the dog. "Any fool can take a great dog and make it greater," Kevin said.

To train people to motivate their dogs, he gets them started thinking inside the box. It was hot that fall day in Seattle. Kevin sat in the minimal shade, and I followed his orders. I knew I wasn't going to be sawed in half or have knives thrown at me. All that was needed for Kevin's magic show were some dogs and handlers and five cardboard boxes in a parallel line on a lawn, three or four feet apart. I looked inside them. I felt around. They were empty. There wasn't a scent source in any of them.

As the magician's assistant, I probably shouldn't betray the magician's secret, but Kevin did give permission. My job as magician's assistant, after putting out the boxes, was to stand there holding dogs, expressionless as a butler. Every single dog—from the blind English spaniel to the squat guy who looked like a puggle (a designer-dog cross between a pug and a beagle) to the pretty little Chesapeake Bay retriever named Truffle—became gullible marks in Kevin George's game of five-box monte.

Kevin, a short, humorous, white-haired, generously proportioned pasha, directed the handlers to make fools of themselves. "If you can't act like a crazy person, you will not be a good dog trainer. Don't be scared to do things that make you interesting to the dog," Kevin told them.

I held the first dog, and the handler started acting insane. She played the role of the monte scammer. She shook the dog's favorite toy in front of his nose. She then ran away, screaming, shaking the toy as if breaking its little neck. She ran to a cardboard box and made a big deal of holding the toy, as if holding a live rabbit by the ears. The dog, the mark in this case, was staring intently—he couldn't help it—as the toy disappeared into the cardboard box, very much like a rabbit going down its hole. The handler then ran back to the dog, who by now was pulling like crazy on my arm. I stood impassively, trying not to wince. I handed over the excited dog, and the handler unclipped the lead. The dog ran to the box with the handler right behind him. Bam. Toy retrieved. Big happy, joy joy.

We did the same thing a second time. The dog was the mark. He knew he couldn't lose. He watched, knowing exactly where the toy was. How cool and how simple. Stupid handler. Stupid handler's helper. On the third run, the game changed. The handler ran to box one and pretended to put the toy in the box. She moved to box two and quietly put the toy in that box. She returned to box one and made a fuss as though the toy were in there. She faked out the dog. Misdirection at its finest.

Kevin again pointed out to everyone that it was critical to act as stupid and obvious as this handler, like a bad melodrama.

Set loose, the dog ran to box one, of course, and found nothing.

The box was contaminated with the dog's scent, the owner's scent, even the toy's scent. But no toy. What betrayal. Puzzlement. Outrage. The dog flipped the box over. Nothing. Then the dog glanced over and saw another box a few feet away. Maybe he got a little whiff of his toy, since the wind was working to help the process. He ran over to the box. Whoa! Look! My toy!

Everyone cheered as if watching a three-card monte game on a street in the Bronx. The dog pranced and shook his toy hard. He was hooked.

"When people get interested, they can get hooked hard-core. They don't like not being successful," Kevin pointed out. "They will put more money down and more money down." The dog, not having much money, put down more and more drive. Interestingly, so did his handler and the others watching. Everyone who was watching was invested. The dog had been scammed. That's what you want. At first.

Kevin folded his hands over his waist and smiled. His favorite part of the scam was coming up. A dog has a huge advantage over a human. Sure, dogs have eyes and ears, but they have noses, too. The dog trying to find his toy soon realized that he couldn't believe his eyes and ears, so he stopped using them. He didn't look at or listen to his screaming, unreliable box-switching handler, asking her to solve the puzzle for him: Please, Mommy? Where's my toy? Instead, as the game progressed, he used what an ancient pre-mammalian ancestor, the shrewlike *Hadrocodium wui,* had bequeathed to him: The dog started thinking and smelling for himself. All the dogs did the same thing. At first gullible and betrayed by their handlers and their own eyes, they started using their noses, working it out for themselves. Soon the dogs were methodically and quickly searching up to eight boxes scattered across the yard, flipping them over with their noses, ignoring their handlers, to get their toys. They were no longer marks. Kevin had gotten the handlers to convince their dogs to commit more and more interest to the hunt.

It's critical that a working dog be able to lead, to independently decide where and how to search, instead of timidly looking to the handler for cues. It is the inverse of the relationship that most trainers suggest we have with our household pets.

I watched about twenty dogs work that day—all shapes, sizes, and personalities. They were hooked; they all wanted the same thing. The big dog-aggressive chocolate Lab who had been getting up in the muzzles of other dogs realized better stuff could be found in the boxes. Dogs? What dogs? Where's my toy?

"The dog itself always has a high degree of interest," Kevin pointed out. "They're very aware of what's going on around them. The simplest thing will draw them in if it's intriguing enough."

As more and more boxes were added, "It should look like fanning cards," said Kevin, spreading his fingers open from a fist, one by one in a fluid movement. This was the cascade of understanding that the dogs experienced as they flowed among the boxes, sorting scent like pros. Kevin's easy hand motion showed the importance of learning things in sequence. And the importance of remembering to keep things magical.

. . .

I didn't know how to keep things magical in my early days with Solo. I was trying to follow basic directions. Just as I started to get my timing down on the buckets—swoop and present, swoop and present—Solo was starting to get bored. Soon he ignored the liver treats. The same thing gets old quickly. Essentially, Solo was learning the parts of speech when he wanted to parse paragraphs. So, of course, did I. But I had read far enough ahead in *Cadaver Dog Handbook* to know that our own desires—to leap ahead of mechanical scent imprinting before Solo had it down solid—were not always desirable.

Nancy had anticipated our boredom. We wouldn't leave the buckets, but she added another layer. She pulled something that was more interesting than liver treats out of her capacious canvas pants pocket.

I took it gingerly. It was a PVC pipe, about two inches in diameter and nine inches long, drilled full of small holes, the ends capped tight with purple plumber's glue. A little bit of death was trapped within on a piece of cloth, its odor gently seeping through the holes. I sniffed, partly to reassure myself that I could do what Nancy had done: stick it in my pocket without having to think twice about it. An old, independent Appalachian woman, increasingly vague with dementia, had wandered away from her cabin. She had been dead twelve days before her family found her. I thought I knew the smell of human decay, having worked in nursing homes throughout my teenage years. But this pipe's smell wasn't the cloying sweetness I remembered. It was a light dry must, like mold on an orange, not potato leaking in the vegetable bin. Just a twist of cloth with dried body fluids provided enough to start training Solo.

It could have been anyone inside that pipe. Cadaver scent is chemically generic, not linked to an individual. Nonetheless, I was secretly glad that my and Solo's introduction to cadaver work was with what I wanted to believe was a benign death. It might not have been. Nancy didn't care. She's not sentimental. She wants to donate her body to the University of Tennessee's Anthropological Research Facility, more fondly known as the "Body Farm," so she can lie out on a hillside, decomposing. Preferably in the open air. Not under a tarp but not naked. Otherwise, she says, she'd prefer to have her body divvied up for search teams to train on.

Solo came to know the pipe as "fish." It signified fun and would ultimately teach Solo that—even more than the buckets—that odor was what he should look for. Nancy showed it to him. "Good fish," she said.

He sniffed. Sure. Whatever. Interesting, but not as interesting as Whiskey. Then Nancy started to gambol about, large and nimble and silly, whipping the pipe around, making it irresistible. She encouraged him to grab it. She pulled it away. He followed, grabbed it, tugged hard, and she let him have it. He won. My, what a strong, big dog you are.

In the world of working dogs, and when laying a foundation you can build on, dogs get to win. They get what they want, they are encouraged to chase it, grab it, chew it, shake it, kill it. One problem is that men are often afraid to gambol around, and even more afraid to lose. Mike Baker, the K9 sergeant at the Durham Police Department, always tells his new, stiff, nervous handlers to raise their deep voices high and get silly: "Come on. Be more exciting than pee on a tree!"

For Solo, the PVC pipe was more exciting even than Whiskey. By the time the pipe was no longer needed to signify that the search was beginning and was bequeathed to another cadaver dog in training, it had fully served its purpose—forever bonding the concept of play to the concept of dead human in Solo's head.

It underwent many iterations. At first fish was the toy that got tossed out, retrieved, played with. Then it became a toy hidden in an easy spot in the yard or house. "Go find the fish!" Then it was the toy that I presented and pretended to throw before tucking it under my arm or in my back pocket, holding out my empty clean hands, sending him forth to find the fish somewhere else. The first few times, he stared at me hard. I know where it is. Your hands aren't clean. Then he entered into the game, yowling before bounding away. After all, it was a lot more exciting than pee on a tree. I was slowly introducing Solo to the idea that he might have to solve a problem on his own, rather than staring at me to try to suss out the answer.

While he was developing his independence, I was starting to appreciate that a dog who uses his teeth to pull Mason jars out of buckets in search of scent wasn't a bad dog. He was just a jackass.

Marcia Koenig, whose German shepherd cadaver dog, Coyote, found dozens of people, still remembers the embarrassment of her trickster dog, born on April Fools' Day, taking the test that FEMA requires for agility and obedience. The little sable shepherd, bored, grabbed the traffic cones set up for her exam and ran away with them. She flunked. Marcia sighed and started back to the laborious task of training her to obey.

Nine days after that, Coyote found the disarticulated bones of a homicide victim. That was her first find but far from her last.

One day, less than two months after Nancy had started Solo and me with our clumsy bucket dance, it happened. Solo channeled his energy. A bit of form emerged from chaos.

I e-mailed Joan, elated: "He was literally running right past the hidden Mason jar with the cheesecloth lid. Scent was seeping slowly through the top. He screeched to a halt from a dead run, his tail went up, he froze, and he turned. It was lovely—it was so clear that his nose said to him, 'that's the smell!' and that his nose stopped him even when his feet wanted to keep going."

6

Distillations

The smell . . . was most closely aligned with musk, yet the impression upon the olfactory organs was more delicate, more subtle.

—A. B. Isham, MD, professor of materia medica, 1875

The moment a person dies, the body starts to decompose, although different cells decompose at different rates. Billions of the trillions of cells in the body that had been chugging along just moments before, mindlessly doing their duty to both conscious and unconscious bodily functions, give up and collectively exhale. The process begins, researchers believe, the moment the heart stops beating. Something incredibly light leaves the body, diffusing almost instantly.

After that, some things go fast and others slowly. Even organs die and decompose at different rates. Hair is one of the last things to go, bones after that, and teeth, which seem so vulnerable while we're alive, hang on forever. This is called the taphonomic process. Bodies become, in the nicest possible terminology, "less organized and more susceptible to outside influences."

All of it is a complex kind of self-digestion, cellular walls rupturing, everything rich with now-unchecked enzymes. Flies know it. They can locate a body minutes after death. Well-trained cadaver dogs know it's the sweet spot.

A decomposing body, especially outdoors, doesn't automatically smell awful. It depends on the days and months the person has been there; what the temperatures, prevailing winds, and humidity have been doing; and what role insects and animals may have played in that body's progression toward dust. I have stood a few feet from a two-month-old body at a human decomposition research facility and gotten a milky, gassy sweetness up my nose, along with hints of leather and bacon. Someone with a touch of synesthesia appropriately called such a scent "yellowish-orange." The next instant, the air vortices shifted slightly, and I smelled nothing at all.

Human decay is more marvelously complex and varied than people claim. It's not a scent version of what Supreme Court Justice Potter Stewart said about pornography: "I know it when I see it."

Except for food, wine, and flora, I don't dwell too much on what I smell. This is not affecting machismo, but something closer to its unbidden opposite: Our brain is hardwired to avoid decay, even Korean kimchi or other fermented delicacies. So while I need to work with training samples without shuddering, I don't spend time wafting them underneath my nose and figuring out their specific bouquets. Inevitably, though, the scent finds its way up my nose. I've smelled enough variants of human decay, or nothing at all, that I would be hard-pressed to narrow the scent down: It's not "unmistakable" and "unforgettable." Those are emotionally comforting adjectives; we want human decay to be unique and special. We want to be something more than another rotting organism when we die.

But there are all sorts of human decomposition pretenders out there, not just in the animal kingdom. In the forests of Sumatra, and now in specialty botanical gardens worldwide, there's a decomposition cross-dresser called the corpse flower. It's one of the largest, ugliest, and smelliest flowers in the world. It blooms once every six years. Its scientific name, *Amorphophallus titanum,* means "huge deformed penis." I'd like to see one, though not smell it. The corpse flower attracts insects, blowflies, flesh flies, and carrion beetles by releasing putrescine and cadaverine. Those chemicals of decomposition are still used for training cadaver dogs, although a number of researchers have established that the chemicals aren't volatile in burials. And they are common. Once you start sniffing, those scents appear everywhere. Some minerals in the red clay of North Carolina vent what smells like death. I've been on more than one futile dig where the crime scene investigators say they clearly smell human decomposition and find nothing.

Although it has been around as long as we have, few researchers know much about the scent of human death. For all the attention that a forensics show like *Bones* receives, and as much as forensic scientists know about maggots, skin slippage, and the thermodynamics of putrefaction, the work on the volatile compounds that vent off human decomposition is just starting. Despite some early medical fascination and training with the smell of death, for most modern scientists and chemists, "odor mortis" is a new frontier. Forensic anthropologist and research scientist Arpad Vass and his colleagues have identified nearly 480 different volatile compounds coming off decomposing bodies. Arpad thinks it will be closer to 1,000 organic compounds (though not all of them volatiles) by the time they are finished creating a DOA database—which in this case stands for "decomposition odor analysis," not "dead on arrival." As Arpad noted with his slightly inflected German accent, the decomposing human is "a pile of pollution."

Why bother sorting out the chemicals from that messy pile? Because ultimately, the knowledge may help create a machine that can help detect the smell of death. It may help scientists develop more effective pseudoscents for dogs to train on.

If it's a new frontier to figure out what compounds are in the air column from human decomposition, it's another stratospheric leap to understand what the dog's nose is picking up from that air column and translating as human remains. No one knows exactly what the dogs are smelling. We can't ask them. Most likely, they are smelling a lot of things mixed together.

"It's a much more complex chemical composition than any other forensic sample, with the possible exception of human scent," said Florida International University analytical chemist Kenneth Furton. He and a group of scientists, trainers, and law enforcement representatives are trying to develop a national set of best practices for detection dogs. Creating a set of best practices is a challenge, and understanding how best to train dogs to detect human decomposition is one of the group's biggest challenges. "There are more knowledge gaps in human-remains detection than anywhere," Furton said.

The things that we humans invented —bombs, manufactured drugs, and land mines—are chemically simple in comparison to our remains. Decomposing humans are not entirely a black hole, though. While no one has yet nailed the perfect formula for what it is that good cadaver dogs think is unique or interesting about dead people—as compared with, for instance, garbage from a suburban household— we do know that solidly trained dogs can tell the difference between dead human, dead deer, and aged goat cheese, or something equally putrefied and odiferous. Arpad noted that cadaver dogs tend to alert on dead sheep more than on other species; he thinks some dogs get confused by the generous dose of sulfur that sheep emit after they die. We emit it, too.

Although sheep may be chemically close to humans, there are, he notes, several significant chemical differences between us. Chemicals we ingest may play a role. Arpad's research laboratory, along with one or two others, found tantalizing evidence that compounds we swallow or inhale—from fluorinated water to asthma inhalants—may play a role in our unique smell after death. It's not much of a stretch to think that our decaying bodies might whisper more than a hint of je ne sais quoi of chemical compounds we've steeped our bodies in during our lifetimes.

"We take in a lot more chemicals than we should," Arpad said. The unanswered question is whether those translate to volatile compounds significant to the dog, even though the carcinogenic compound carbon tetrachloride seems to jump out when Arpad takes air measurements of human decomposition.

We have a fair amount of evidence that dogs do fine at detecting remains that are hundreds of years old—long before fluorocarbons and freons and fluorinated compounds and solvents and poisonous cleaners and antibiotics came on the scene. In rural areas where water isn't fluorinated, dogs can find people who have mostly drunk well water all their lives. Domesticated pets ingest plenty of fluorinated city water and chemicals, and well-trained dogs don't alert on their remains.

Arpad believes they are getting close to knowing what dogs find important and unique about human remains. The chemical portrait gets blurrier with buried bodies, which may be emitting different volatiles than surface ones. He thinks he knows what volatiles dogs are alerting on in burials—perhaps as few as thirty compounds—but he's not yet tested the theory. Those volatile gases have to be available at the soil surface for the dog's nose to collect. Even bone has twelve detectable volatile compounds.

It is possible that a unique volatile compound—something that barely registers on the gas chromatographs positioned around bodies in Arpad and his colleagues' experimental research plots—hits the dog's nose and lights up her brain like a pinball machine. Or it could be a few compounds, or a bunch of them in various delightful combinations. When you combine the variety of conditions under which dogs find human remains, from freshly deceased to hundreds of years old, the options become dizzying. For instance, Arpad notes, the odor profile for a body that has all the busy microbes associated with the digestive system working overtime is very different from the profile for scattered limbs.

While we don't know the exact compounds they are smelling, dogs find human remains. That should count for something. Just because dogs find the dead without an utterly coherent scientific theory as to why, and without an easy way to test it under controls, doesn't mean it's not happening.

Cadaver dogs' ability to find human remains may be analogous to humans' ability to recognize faces. Researchers don't have a clear understanding of how facial recognition works, but people do it effortlessly, at all angles, even in bad light. Trying to break that recognition down into parts (it's those great cheekbones!) doesn't work. It's the whole face. Machines still aren't nearly as adept at facial recognition.

So the two scientific arenas—what volatiles in what percentages are involved in human decomposition itself, versus what volatiles dogs react to as cadaver scent—may overlap a great deal. Or not. Who cares? Well, we like to know things. It might help us figure out why some dogs are drawn to rotting trees like sycamores or to the vegetative decomposition in swamps. Isolating what the dogs find significant in decomposing humans might help handlers and trainers identify which training aids are best and the ideal way to store them. Ultimately, that knowledge might lead to finding substitute training aids that are legal, safe, and much closer to the real thing than the pseudoscents or other chemicals in current use.

This new research arena has started to upset old beliefs. Some trainers and handlers—more in the past but a few in the present—have claimed that pig samples are the next best thing to human samples for cadaver-dog training. The temptation is great. Pigs are plentiful and a forensic fallback, a model for studying human disease and decay. No ethical fuss or muss about using them as training aids: Just pick up some pork spareribs from the grocery meat counter. Yet, as Arpad Vass noted, there's a vast difference between the chemical profiles of deceased humans and swine. "Pigs smell quite different from humans. We have this well documented."

Nancy Hook scoffed from the beginning of training about handlers using pig tissue as training samples. It was pretty simple to her: "Pigs aren't people."

Mary Cablk, an associate professor at the Desert Research Institute in Nevada who trains cadaver dogs and does research on the reliability of detector dogs, took the pig problem a step further. She and her analytical chemist husband, John Sagebiel at University of Nevada, Reno, compared the volatile organic compounds of chicken, cow, and pig with those from human remains. Their results should be the final nail in the coffin for an entire cadre of trainers and handlers who have said for decades that pig samples are the next best thing to human samples for training. Their research shows that we humans smell much more like chicken than pig when we decompose. If it's any comfort, we smell like organic chicken from Whole Foods.

. . .

. . . and suddenly, coincident with death the beam end dropped with an audible stroke hitting against the lower limiting bar. . . . The loss was ascertained to be three-fourths of an ounce.

—Duncan McDougall, MD, "Hypothesis Concerning Soul Substance," *Journal of the American Society for Psychical Research,* 1907

Cadaver-dog handlers and trainers have watched dogs find human remains for several decades. No matter. No rigorous scientific studies had shown how well dogs could do it. They just did it: Dogs were indicating that they smelled human decomposition. No one knew exactly what they were alerting on, unless a body, or a part of a body, was there as undeniable proof. What part of the part, though, were they alerting on? How soon after death could dogs detect something? How long after death did the scent last? And what, for a dog, was that "scent"?

Despite numerous studies of how working dogs reliably detect drugs and bombs, few scientific studies have featured cadaver dogs. It's messy work, and the real-world aspect can irritate scientists in search of conclusive data. It's hard enough figuring out how a narcotics dog can find pure heroin; figuring out how cadaver dogs can find the infinite variables of dead humans seems impossible.

Scientific uncertainty hasn't kept people from using dogs world-wide to help pinpoint the perfume of death, from the faintest emanation off a tooth to the fulsome scent that lofts from an entire body. A few small studies existed here and there, like Debra Komar's 1999 study on cadaver dogs' ability to find scattered human remains in Canada, and another in 2003 on dogs' finding buried remains in the southeastern United States. That 2003 study noted quite accurately that "dog handlers affected the reliability of the cadaver dog results."

It was far past time, at the beginning of the twenty-first century, for someone to come up with a controlled scientific study on cadaver dogs' capacity to detect the faintest smell of death. That was when an unfortunate series of events—almost always the case when cadaver dogs are involved—created a perfect opportunity for imagining, then realizing, a small, elegant scientific study.

A wife in Hamburg, Germany, disappeared off a yacht in 2005 while sailing, apparently not entirely happily, with her husband. The husband reported her missing. She must have fallen overboard, he said. When the Hamburg state police sent one of their cadaver dogs onto the boat, the Belgian Malinois aggressively communicated to its handler that something bad had happened on the mattress in the yacht's bedroom.

There was no body in the bed. No blood. No tissue. It must have been a soupçon of scent that caught the Malinois's nose. Who knows? Mattresses can be funky places, even on the best yacht. Regardless, without a body, tissue, or blood, there was no case.

The prosecutor, however, wanted to believe the dog. He contacted Lars Oesterhelweg, a forensic pathologist then at the Institute of Legal Medicine in Hamburg, and asked him to set up a study—not replicating the entire yacht, but providing more definitive proof that dogs can scent death without the presence of specific forensic evidence, like blood, tissue, or bone.

For his study, Oesterhelweg and his colleagues used two recently deceased men, A and B, sixty and sixty-three years old. They had collapsed and died on the streets of Hamburg. Sometime before, they had agreed to donate their bodies to medical science. A and B could not have known how delightfully and noninvasively their bodies would be used. Each freshly dead man was whisked into the local hospital's inner courtyard, wrapped in a cotton blanket, and laid, for just two minutes and ten minutes each, on top of brand-new carpet squares that sat on new tables in the open-air courtyard. Oesterhelweg didn't want a hint of hospital contamination. The carpet squares received nothing but that indirect exposure to two-hours-old bodies: no tissue, no fluid, no blood, no rafting cells, no DNA. My romantic, noninterventionist side likes to imagine that was the beginning and end of A's and B's sacrifice to medical science: this gentle wrapping, this brief application of their bodies to carpet squares. But that's neither the reality of good cadaver use nor good recycling.

For the next two months, three Hamburg police cadaver dogs were asked to show their handlers which carpet square, among a group of uncontaminated carpet squares, contained cadaver scent. The most mysterious fragrance on earth was no mystery to the dogs. Two of the dogs, B and L, were almost 100 percent accurate. K scored 90 percent. That's the reality of working dogs. A few are excellent, while some are very good. (Others are horrid. The last category wasn't represented in the study.) The small study's results: Well-trained cadaver dogs can smell the faint remnants of the odor of death, impregnated on a carpet swatch, for months following the brief presence of a newly dead person.

The study showed, Oesterhelweg wrote in what amounted to scientific effusion, that well-trained cadaver dogs are "an outstanding tool for law enforcement." I tried to find out, to no avail, if the prosecutor had pursued the case. Never before had I wanted so much to be able to communicate in German.

Every study has its limitations. This one didn't answer the question: What exactly were the dogs smelling in those carpet squares? They could alert reliably, according to the study, but what was in that fresh perfume of death?

Arpad wistfully asked me if the German forensic scientists had done a headspace study on that early perfume. Did they measure what volatiles were in the airspace in the containers that held the carpet? I doubted it, but I could see why Arpad wanted the information. By the time donated bodies get shipped to the University of Tennessee Anthropological Research Facility, their decomposition is much further advanced. In the seconds and minutes after death, Arpad thinks it's possible that compounds such as ammonia, hydrogen, carbon dioxide, and methane escape from the body. Such compounds don't weigh much at all.

They are lighter than air.

. . .

People are spending more time worrying about the possible nuances of training than going out and training their damn dog. Our dogs are very forgiving, so you want to try working on some source. See what your dog does. It's not a big frigging mystery.

—Andy Rebmann, 2012

Every training aid for every working dog has its drawbacks. If you are training a bed-bug dog, you must find a way to keep your bed bugs fed and happy (and contained). If you are training a bomb dog, you have to be comfortable carrying nitroglycerin and gunpowder around in your trunk.

If you have a cadaver dog, you run into a different kind of supply and storage problem. Having diverse materials to train on is crucial, and "decomp," as it's called, comes in all varieties: from recent blood on a surgical swab, to lovingly harvested dirt from a Civil War burial ground, to six-hundred-year-old bones from the Mississippi Delta.

Happily for Solo, North Carolina has reasonable laws about cadaver-dog training material. The wisdom tooth of Nancy's ex-husband was only the beginning. Solo would need a variety of materials to train on, from fresh to older to ancient. Material that would, albeit at a scaled-back level, mimic what he might find out in the woods, buried, or in an abandoned building. So in the midst of answering student e-mails in my university office, I'd find myself distracted, mail-ordering a search-and-rescue dog package from the Bone Room in Berkeley, California—although I called someone at the state medical examiner's office first to make sure it was legal. It was. I was delighted when the order arrived in the mail: a small box with an ivory tibia and some toe bones in a plastic bag. The foot bones looked remarkably like the Styrofoam peanuts they came nestled in.

Less than a year later, evidence started mounting that some of the plastinated bodies displayed in the popular museum show *Bodies: The Exhibition* came from prisons in China. Curious, I started researching the possible source of my specimens from the Bone Room. The results were inconclusive, but I realized that I needed to find other sources. So, obviously, did the Bone Room, which posted on its website: "We regret to inform you that our stock of damaged and discolored bones, set aside as Search & Rescue Dog training packages, has finally been totally decimated, and we will not be able to offer them for a while. Unfortunately there are absolutely no human bones coming into the country at this time, and while we are currently trying to find a source that would hopefully include more of the 'dog quality' specimens, we do not yet have a line on a possible exporter."

When having my teeth cleaned, I talked with my dentist about how to get hold of a couple of molars for training. He was happy to provide a few, and kind enough to lend me, briefly, the jawbone from an anatomical teaching skull that had been in his office for decades, to see how Solo would behave. Solo reacted, though he was clearly surprised when he put his nose next to it. He was underwhelmed by the source: That's it? I reacted to that bit of scent? I, on the other hand, was thrilled and reported the outcome to the dentist when I returned the jawbone.

A K9 officer from a nearby city provided me with carpet from a processed suicide scene. One warm summer evening found me with a friendly death investigator who had handled cadaver dogs. We were in the parking lot of the police K9 unit's headquarters, cutting up material for training aids: an old sheet that had lain for days underneath an undiscovered body in an apartment. It was malodorous but not nauseating, and we didn't need the whole thing: just six-inch-by-six-inch patches that we could pop into Mason jars. The scissors would be bagged and tossed. We knelt in front of our bounty to make careful cuts, then rocked back on our heels. We had to make sure our double-nitrile-gloved hands didn't touch anything except the material. We were smiling. Solo was whining in the car, waiting for training. It was a beautiful evening.

I permanently borrowed a wide variety of kitchen storage containers from David, whose home-chef habits benefited me on several fronts: Mason jars, honey and jam jars, bigger Rubbermaid and Cambro and gallon glass jars for holding the smaller jars, all storage for Solo's training materials. I bought others new. Using an assortment would prevent Solo from associating the smell of any one container with the scent of the dead. If I were ever called to testify, I wasn't going to be trapped by a good attorney into admitting it was possible that my dog was trained to smell and alert on plastic storage containers, or Ziploc bags, or cotton sheets, instead of the human remains I was trying to train him on.

There's great debate about whether dogs get rewarded accidentally for finding whatever storage material you are using instead of the training material you want him to find. Or for alerting on the smell of the gloves you use to place the samples, for that matter. Some trainers and handlers contend that material should be stored only in glass jars with metal lids, or it will be contaminated, and you will inadvertently cross-train your dog on plastic bags. Others insist that material needs to be stored in everything and everywhere: teeth in the freezer in a freezer bag, dried blood in the fridge in a jar, bone in the pantry.

Almost everyone believes that if your state laws allow it, it's ideal to train your dog on the wide range of stages humans go through before they disappear entirely—from fresh tissue and blood, to what's called greasy or wet bone, to adipocere, to the dry bones of the desert, and even the ashes of the crematorium.

Everyone agrees that exposing dogs to whole bodies, in their various stages of decomposition, is great training. A cadaver-dog handler's version of paradise is tucked into the western corner of North Carolina. Western Carolina University calls it the FOREST, an acronym for Forensic Osteology Research Station. Do not, even lovingly, call it a body farm in front of its founders. The site is up a gravel road and a wooded hill, surrounded by a huge cyclone fence with razor wire.

Paul Martin recently graduated from Western Carolina with an anthropology degree. Paul's research work kept creeping closer to his dog work. He was a former sheriff K9 handler, then a cadaver-dog handler. As an undergraduate student, Paul realized that the new FOREST could help not only forensic anthropologists and their students with training and research, but also cadaver dogs and their handlers. Though Paul has moved on to a graduate program at the University of Mississippi, he helped found and continues to organize cadaver-dog seminars at WCU. Dogs and handlers get a chance to train in the small fifty-eight by fifty-eight-foot plot. It might seem an odd highlight for a seminar, but serious handlers know how crucial it is that both they and their dogs accustom themselves to what they might encounter out in the world.

Dogs used to training on small samples, as most samples are, can be overwhelmed by the large pool of scent an entire body can emit.

Bodies on the ground, even live ones—as everyone knows who has played with dogs and puppies by lying on the ground—can be scary. I've watched many dogs react at the FOREST. Some come up to bodies with great hesitation, sniffing cautiously, then starting back. The sound of flies or maggots moving inside a bloated body can freak them out. A few growl. Some crouch and crawl up to a body. A few bound up to the bodies in good humor. That's great, but a handler worried about an enthusiastic dive into unctuous remains can jerk back the lead so quickly that she corrects the dog when she should be rewarding that lack of fear—while still keeping the dog from doing damage.

Paul talks quietly to all of them, keeping an eye on the dogs and handlers, making sure that no dog does a belly flop into the remains, but that none is discouraged with a too-harsh leash correction. He soothes both handlers and dogs as they negotiate the small plot that has ten bodies in various stages of decomposition, from a body bloated like the Michelin man to skeletonized remains to buried ones. Enthusiastic or fearful dogs pull handlers in all directions while the handlers work on controlling their dogs, praising them, and keeping their own balance.

"She's doing good," Paul told one handler, and then burst out crooning, "Good dog! Good dog! Good dog!" as the border collie decided it might be fun after all and went straight up to a body, wagging her tail.

That kind of hugely positive experience can set up a dog for a happy life finding the dead.

· · ·

As much as I was learning about the scent of death, it would be some time before Solo was exposed to a whole body. These were early days of training for us. Solo was ten months old. We'd begun training with Nancy five months before. By this time, January 2005, I had started to carry around cadaver training material in a small picnic cooler.

David and I were headed to the beach with friends and dogs piled into a rented SUV. I wasn't dreaming about romantic walks on the beach at sunset. I had started seeing the North Carolina landscape as one endless opportunity to train Solo. As David drove, I stared out the window at loblolly pine plantations and abandoned concrete-block buildings. Could we train there? That long-harvested soy field looked like a promising place to stash a cadaver hide. Sand, bitter blustery wind, and prickly pear at Kill Devil Hills sounded like a great way to challenge Solo in a new environment.

Solo learned to find his "hides" at the beach, in the garbage can at the end of the driveway of our rental house, in the backyard sand, amid a patch of prickly pear. I was elated as I wrote my training report. On the second night at the beach, I called Oregon to talk with Dad about our impromptu vacation. I told him I would send him pictures of his Megan, who had been floating blissfully up and down the beaches, a maroon-colored wraith in the winter fog. Dad sounded dreadful, his voice thick and slow. His hip had been hurting the last six months, since shortly after he had visited us in North Carolina. While we thought it was a side effect of Lipitor, it hadn't improved. Only good Scotch, Dad said, made it feel better. I got off the phone and cried.

I had never said much to Dad about what I was doing with Solo; I didn't know why, exactly. I couldn't talk to him about bodies and crime—it felt base, rather than biological. Part of it was that he was an academic. What I was doing with Solo wasn't academic, much as I might contend that it was about reading winds and understanding decomposition and scent patterns. Part of it was that he didn't fully understand my odd love of dogs like Solo, rather than the female setters who draped themselves on him, placing their paws on the sleeves of his old cardigans, pulling the threads out ever so gently and insistently with their untrimmed nails.

Perhaps somewhere deep down, I knew. When he called a few days later, we were home from the beach. I was oddly unsurprised that Dad finally had a diagnosis. I was shocked at how far along it was. Cancer specialists, I have learned since, always say the same thing to Stage IV patients: You have six months to a year. Even when you don't.

Training Solo stopped. I could not bear thinking about death all the time. In any case, there was no time. I left David alone with Solo and Megan and flew back and forth to Oregon to spend a couple of precious remaining weeks with Dad.

He was dead just seven weeks after that phone call. He was cremated in the Pendleton wool bathrobe he had loved so much. We cast his ashes to the winds in the Cascade Mountain meadows of Oregon near his home.

7

A Spare Rib

When the Man waked up he said, "What is Wild Dog doing here?" And the Woman said, "His name is not Wild Dog anymore, but the First Friend, because he will be our friend for always and always and always."

—Rudyard Kipling, *Just So Stories,* 1902

I didn't dream of Dad at first. In that vaporous line between wake and sleep, though, I would relive his last weeks and sit up, pulling in my breath.

Everything and nothing had changed in those few months. Solo now looked like a powerful Velociraptor—his head still too big for his body, his tail like a large motile rudder. He'd missed me as I flew in and out of town; he let me know by screaming and crying, then dashing off to grab a toy to shove in my lap: Play *now*. He wasn't a dog who did depression. If I stared at him vacantly, that was a cue to play fetch, not to comfort me. Solo had bonded even more with David while I flew from coast to coast. While Solo's so-called obedience training might have suffered slightly in my absence, he and Megan were now compatriots. They spent their time on a Kabuki-theater version of dog play: all performance, dance, and tease. Megan didn't believe in contact sports. In my absence, she had trained Solo to interact in a way I couldn't have imagined him capable of with another dog: playful, subtle, light on his feet. He still swaggered, but he had developed a sense of humor with her. He needed more time outside, where he could run and sniff and not be in my face. With me, he thought contact sport was required.

I thought about the options. I could continue to mope. Or I could call Nancy Hook. I called Nancy. Parents' deaths create a void, she said. No matter what you thought of them. And you really liked your father. Come on out and bring Solo to train. So I did.

I had missed an entire season at Nancy's farm in Zebulon. I left while it wore the tans and grays and browns of early winter, though undershoots of brightness lurk year-round in North Carolina. Nancy had exchanged her winter Carhartts for lighter camouflage-green pants stuffed into tall rubber boots. Her blond-red hair was stuffed under a baseball cap instead of a wool knit cap. She had become one of my measuring tools of normalcy—her laughter, her ease in the world, her ability to be simultaneously direct and comforting. Her ability to slap me upside the head without it hurting too much. She was sane, and her sanity infected me.

A well-meaning friend asked me, "Are you grieving properly?" Probably not. It was hard to grieve in the middle of a soft Carolina

April, walking through muddy chartreuse fields filled with mist and cow pies. And cows. It was hard to grieve when I had to watch my feet and control Solo, keeping him away from Whiskey's fence, from the chickens doing their herky-jerk insect dance in the yard, from Rocky, Nancy's Morgan horse who didn't suffer dog fools gladly, from the Herefords looking doleful, then dour.

"Get that high, panicky tone out of your voice," Nancy told me as Solo started lowering his body toward the ground and giving the eye, preparing to stalk the mama cows with their new calves. I dialed it down a notch and used my low big-girl voice. It worked. Solo reluctantly swung back toward me, and I hitched him up until we got farther from the cows, and I could free him. He wanted to make clear that he was the winner of the dog–cow debate, so he sauntered over to lift his leg on an electric fence before he started his search. We could hear the *bzzt* of electricity from twenty yards away. He didn't flinch. That, Nancy told me, is exactly what you want in a cadaver dog. If he could ignore that jolt, in that part of his anatomy, nothing would shut him down.

I didn't shut down, either. I trained Solo several times a week. I found new places to train. At our local feed-and-seed warehouse, where pigeon poop and running mice distracted Solo only momentarily. At the mall across the street from our home, where new construction materials—including pallets and concrete blocks and pipes—created a scaled-up version of our early foundation work. Check here, check here, check here.

Solo created new problems for me to think through. He was a wild dog. My training notes were littered with issues: He needed a better alert. He was distracted. We'd transferred to a down-and-whine alert, but he wouldn't whine. Then he wouldn't down. When we transferred to a rubber Kong toy for a reward, he didn't want to give it back. He didn't want to search next to the dog lot if Whiskey were there and the two of them could fence-fight instead. He would stare at the cows and wonder: What merriment could he pursue? Nancy reminded me to keep my voice low and forceful.

Solo's puppyhood issue of being a singleton kept raising its ugly head. One day during training, an obsequious stray—a female yellow Labrador—came running and wagging and crawling across the field while we were training. Solo flashed his teeth, rolling her repeatedly. Then he attacked Wolfie, the German shepherd search dog he should have been collaborating with. Except for Megan and two or three neighborhood exceptions, he hated dogs his own size. And small dogs. They were a pain. Nancy rescued Boston terriers, and I had to rescue one from Solo. Solo lunged, the grass was wet, my foot slipped, my voice skyrocketed into high soprano. These were the things that made me cry and made Nancy shrug. Solo hadn't killed Yankee; he'd just considered it.

"Blast, blast, blast," Joan wrote me after I described yet another incident of unfortunate behavior. "Aggression and the canine mind are so very interesting . . . and sad, when it is your dog. The one thing I can say is that if he wanted to do harm, he could and would have. So, as nasty as these incidents are, he doesn't appear to be hurting dogs physically."

She was right. He was all teeth and hackles and growls. He never drew blood. Nonetheless, David and I made a difficult decision and neutered him, hoping a touch less testosterone would move the needle on his tachometer down a few points. Afterward, I threw him back into obedience classes. I learned to time interventions, to break his stare-down with another dog by blocking with my body, to de-escalate, to communicate a clear message: Obey no matter what, you little shit.

We survived the Night of the Snapping Terrier without even a growl on Solo's part. We got through the happy-pit-bull-adolescent-on-top-of-the-shepherd incident without tooth hitting skin. We realized that Solo would never be normal with other dogs. Yet he was becoming more and more responsive with us. He was no less energetic, but he was becoming our friend. He would look at us steadily with his dark chestnut eyes and occasionally even lay his heavy tadpole head on our laps and fall asleep without demanding instant game gratification. He loved humans, including babies and children.

Nancy was making our training problems harder and harder. Before I knew it, it was midsummer, less than four months after Dad's death. Solo was fifteen months old. It was easy for me to keep track of Solo's life and Dad's death: They ran on parallel tracks, going at the same speed, but heading in opposite directions. Solo's first birthday and Dad's death were five days apart. Then there would be a second birthday for Solo and a first anniversary of Dad's death. On it would go.

Solo and I stood in a large cow pasture north of Nancy's house. Great blue skimmers and common whitetail dragonflies, looking like pieces of chalk on the wing, buzzed across clumps of cut grass to land on cow pies. It was humid and hot. Solo and I stood at the top of the pasture. It had a swamp and pond at the bottom, where some of the cows hung out. Solo was whining as I held him. Before I sent him, I tried to figure out which direction the wind was coming from; the moist air was barely eddying.

"Find the fish!" As Solo pushed away from me, Nancy narrowed her eyes and looked at me, giving me the international symbol to shut my mouth and keep it that way. She threw away the invisible key. Yes, sensei. No nervous chatter, outside or inside. I channeled Bruce Lee. "A good fight should be like a small play, but played seriously. A good martial artist does not become tense, but ready. Not thinking, yet not dreaming. Ready for whatever may come."

The field was huge. I'd worked it before with Solo, but Nancy had conveniently started forgetting where she had planted some of the training materials. She was also teaching me how to sketch a search pattern by fixing on spots on the horizon. I told her that I'd use the big deciduous tree on the hill as one marker. She laughed at my highbrow botanical vocabulary. Solo ignored my pattern. He threw his massive head up and ran downhill, into the cow-poop-filled swamp. He slowed, his tail stiffened into the loop that let me know he was near cadaver material, and he lay down in the muck, staring at me, silent. His new alert. We'd abandoned the whine as an alert two months back, when it became increasingly clear to Nancy that his life was one big whine. We'd abandoned the food because a Kong on a rope was more fun.

"Throw it. Quick!"

I obeyed, clumsily.

Nancy gathered up the training aid. There was more out there. She swept her hand up and across the pasture. Anywhere out there. It was hotter out. Solo panted, not getting a whiff of anything. Nancy critiqued my pattern. Too much zig. Not enough zag. I wanted to give Solo some water out of my new water bag. She reminded me that we had been working for less than twenty minutes and it wasn't that hot out. I gave him some water anyway. It gave me a chance to catch my breath.

I restarted him. Solo's head went down. He slowed even more, plunging his nose deep along a high ridge of grass. Then he moved away, gaining distance along the ridge, about thirty feet straight away from me. He circled and stiffened. Then he was down, toenails dug hard into the ground. Bam. Head back to me, brown eyes fixed. This time I decided to trust him. He had been so clear. I moved fast and flung the Kong. Solo growled and yowled and tossed the Kong for himself, bouncing it off his nose, rapturous.

Nancy walked over, lifted the dried cow patty, twice the size of a dinner plate, and showed me the prize: a few inches of desiccated bone that looked like a small beef sparerib, a donation from a friend and fellow dog trainer. His own rib, removed in a surgery.

She looked at Solo. She looked at me. "Damn," she said.

Names are not always what they seem. The common Welsh name Bzjxxll-wcp is pronounced Jackson.

—Mark Twain, 1897

"Napoo," the handler softly told the huge bloodhound. "Find napoo." The ginger bloodhound sauntered off toward the underbrush, her ears and body swaying, brow wrinkled.

I'd started training with a group of handlers from the foothills of North Carolina. I'd get up at four-thirty A.M. to be on the road by six, just as pale streaks started to lighten the horizon, driving on back roads through Mebane and on to Reidsville, the center of the American Tobacco Company until the mid-1980s. I learned to appreciate country ham and biscuits and Dunkin' Donuts coffee at dawn, surrounded by baying bloodhounds and men in camouflage who chewed or smoked. I liked them a lot. They tolerated me, and they taught me. For nearly a year, I didn't tell the dog handlers much about my work as a professor or my politics, even when we were eating lunch together at a Golden Corral. Nor did I tell most of my university colleagues about my weekend walkabouts.

Ken Young—with his military bearing, his trimmed mustache with a sly smile beneath, an olive fatigue cap, and a pistol strapped at his side—ran a florist shop. On weekends, he ran dogs and people. He would stand in front of a group of slouching handlers in the firehouse, many of them with cuds of tobacco tucked under their lower lips. Ken's version of the classic sign-off that *Hill Street Blues* sergeant Phil Esterhaus gave the gathered day shift, "Let's be careful out there," was "Now, let's go have some fun out there."

We did. Nancy and I early on settled on the command "Find the fish" for Solo. The cadaver-dog handlers from the Piedmont foothills told their dogs to find napoo. They told me it was a Navajo term for the dead. It seemed to have spread far beyond the Southwest, going as far north as Canada and as far east as North Carolina. It was an evocative and mysterious command. Some handlers added "ka-ha" to the front of it, as in "ka-ha napoo."

The unusual command, handlers said, kept family members on the scene of a disappearance from getting more upset. If there were reporters around, they wouldn't be clued in that the dogs were searching for a body. Both claims seemed a bit of a stretch, especially since the media appear entirely clued in, if there, and families, while ever hopeful, mostly aren't idiots.

One Pennsylvania reporter explained about the supposedly Native American command, "Because it is in a foreign language, the command won't be confused with other speech."

"An instructor uses an Indian word for aesthetic reasons," another newspaper reporter wrote of cadaver-dog training.

The term isn't Native American, and its provenance isn't particularly aesthetic: "Napoo" is British and Australian slang from World War I, a bastardization of the French expression "Il n'y en a plus [*ny-an-a-pu*]." No more. All gone. Finished. There is no more of it. British and Australian soldiers became accustomed to two things in France: no supplies on shop shelves, and death. They used the term "napoo" to cover everything from that Gallic shopkeeper shrug that says, "Sorry, but we're out of beer," to death in the muddy trenches. "Half the platoon got napoo'd last night."

We humans love words and the stories they tell when they get strung together. It doesn't matter if they're true. Those who would trace the provenance of dog commands are certain where they originated.

"You could tell who people were trained with," said one seasoned handler in the Northeast. " 'Find Fred.' That was Andy's." Another handler from another part of the country told me, shaking her head, that she always thought Andy Rebmann's favorite command, "Find Fred," was deeply insensitive, especially if family members were on the scene.

The only problem with the story about Andy's command is that, just like the origins of "napoo," it doesn't appear to be true. Andy shook his head, though his eyes glinted in sardonic amusement. Never, he said, had he asked a dog to "find Fred."

"I always use 'Look for it,'" he said.

Darn it. "Look for it" is so prosaic. Nor does it capture the essence of brutally practical and politically incorrect Andy the way "Find Fred" does.

"Mor-te," a North Carolina handler tells her big German shepherd, with an emphasis on the T, so the word ends up with two syllables.

"Where's Mortimer?" another handler urges. "Where's Mort?"

"Where's Chucky?" a handler asks her border collie.

"Find bones," Marcia Koenig, Andy Rebmann's wife, tells her German shepherds. That one seems to work just fine. Marcia's dogs have found dozens of bones over the past two decades.

The best term, however, belongs to Suzi Goodhope. "Hoffa," she tells Shiraz, one of her Belgian Malinois.

Not only was Solo learning, and not only was I learning to handle Solo, I was learning an entire new vocabulary, listening to the arguments that would crop up: Should we call that behavior an "alert" or a "final indication"? As long as it's clear in your records, it doesn't matter.

Andy was on his cell phone with Jim Suffolk, and they were reminiscing together. "We used to call them body dogs," Andy complained. "Now they're calling them HRD dogs. Human-remains detection." Actually, Andy was leaving out the newest FBI-approved term: "victim-recovery canine." Or, more obliquely, "VR canine."

Language always seems to move in a direction where specificity gets lost. Bodies, and death itself, start to disappear under the weight of those terms—collateral damage.

Roy Ferguson out of Sevierville, Tennessee, performs a quiet ritual he started with his first cadaver dog, Cherokee. He sits his current dog, Apache, at his side, and gives the slight black-and-red German shepherd a sip of water. It clears his nostrils and mind. To a single-purpose dog, those rituals are more crucial than any command. When Roy says "Zuk Morte," Apache already knows what the game is.

Solo was a fast learner. He was now cued in to the pants and boots and even socks that I wore for training, ready to go as I dragged them out of the dryer, pulled them on, stumbled downstairs. As David made coffee and I brought in the newspapers, Solo would dash from the bedroom to the front door, from the yard to the car, and then back to the house. The gray light was turning pearly; the sun would be rising soon. Stop grieving. Stop dawdling. Outside is better than inside. Let's go.

8

Comfort Me with Bite Work

"It says, madam," the maître d'hôtel went on, "that you are over your limit."
He leaned down and hissed menacingly. "Do you know what your limit is?"
—Ruth Reichl, *Comfort Me with Apples:*
More Adventures at the Table, 2001

Making dinner for good friends was one of the greatest pleasures in our
lives, a familiar regimen for David and me. The anniversary of Dad's
death had come and gone, I was turning fifty, and habit again dictated
our days. We were back to normal, but it no longer felt as comforting

as it had. Time to bake the bread. Time to spread out the spread. I should have been grateful, but I wasn't. Food and friends were essential but not sufficient.

I was no longer comforted with apples, with flagons of wine. I was sick of routine, at the university and at home. I no longer wanted to pretend an undying devotion to academic life, despite its many privileges, despite my love of teaching—and despite my worrying about how my colleagues might perceive my dogged wandering and wondering. The need to pretend that the world of theory was my natural home had disappeared with Dad's death.

I had lived for half a century. David and I, if we were lucky, might be able to squeeze in a couple more decades together. If he didn't lose his patience with me.

"I just want to be happier," I told him. I could hear both the self-indulgence and the sloppy thinking in the demand. I needed to dial back on the whine. Some years before, a friend and university colleague asked, "Why can't you just be happy?" I tried to answer her honestly. Happiness was not something I aspired to. I had a good man, a good job, and a dog who was trying to be good—and I was left wanting more. I wanted to be challenged and engaged, to fire on all six cylinders. I wanted to do something that made me lose track of time. I wanted to push my limits. David was a master cook and baker and had apprenticed with one of Paris's best artisanal boulangers. He hung out with chefs, thought about food, and wrote about food. I wanted to master handling scent-detection dogs.

Training Solo to do search work, researching the science of scent, and spending my time with working-dog people was the closest I had gotten to that desired sense of immersion in more than a decade. David and I had been pretty immersed in each other for the first few years of our relationship, but our marriage now gave me the space to play. I kept getting drawn to North Carolina woods and fields, and the dogs and dog people who inhabited them. I was starting to admit to myself that nose work and research were becoming more engross-

ing and fascinating than my academic research and writing projects. I wanted to do something practical in my spare time, something that engaged two neglected parts of my inner 4-H child: my heart and hands. My head and health got plenty of attention.

On my birthday night, after our friends departed, David and I talked while both dogs slept. They were exhausted from showing off for company. I knew that taking my and Solo's training to the next level over the coming year would throw our comfortable routine out the window. My midlife crisis would not be assuaged by buying a sports car. David, who loved routine, knew that, yet he was willing to come along for what was bound to be an unpredictable ride.

If I were ever going to use Solo in an actual search, it was time to take the leap: getting us certified to do the work. Certifying Solo meant I would be propelled back into my student years of test taking rather than test giving. Nancy Hook made it clear that certification wasn't enough. I needed to find experienced trainers to work with, beyond her and the Piedmont bloodhound team. If I were ever going to deploy Solo, I needed to up my game, to "bond with the badge," that is, to find a way to work with law enforcement. If the first ambition, certifying Solo, was uncertain, the second ambition—getting access to the world of law enforcement—was a brass ring.

Cops and courts weren't alien territory to me. As a newspaper reporter, I had covered crime and law for years. Several detectives had asked if I'd ever considered entering law enforcement. I had thought about law school, but I'm not sure what side I would have landed on: prosecution or defense. Ambivalence defined my relationship with any number of institutions, though, including the places that cut my monthly paychecks. I was simply adding more ambivalence to the mix. It goes both ways. Law enforcement is reasonably and notoriously suspicious of volunteer dog handlers. We can be an eager lot, with more verve than sense, overly sentimental about dogs, naive about the complexities of the law.

Lisa Mayhew, who investigates children's deaths across the state for

the medical examiner's office, wanted me to take the next step, too. She had handled cadaver dogs and knew a great K9 trainer who happened to work in my city. It took me an hour to compose the e-mail to Sergeant Mike Baker, head of the K9 unit for the Durham Police Department, asking whether I could meet with him. It was less than a month after my birthday epiphany.

"Sgt. Baker," my e-mail began. The hierarchy of academic life made the title easy for me to use. Paramilitary organizations have nothing on higher education. Despite the perception that English departments are bastions of leftist anarchy, we have as many ranks, even more forms and reports to fill out, and a stricter definition of who counts and who doesn't, from adjunct lecturers to vice provosts. I made my tone formal, respectful. No humor. No mention of my academic life. Sure, I was an associate professor at a land-grant university, but in the world I wanted to enter, no one cared. On the contrary, it would be best to avoid mentioning it.

Solo and I, I wrote, "have been training fairly steadily since he was eight months old." "Fairly steadily" was fairly accurate. I had set Solo's training aside for just three months when Dad died. It had felt like so much longer.

I waited. Five days.

Mike Baker e-mailed back, apologizing for the delay, since he'd been out of town. "Hi, Cat. My specialty is in law enforcement dogs, but many of the same theories/principles apply to what you do. If nothing else, I'm sure we could expose Solo to new training areas and obstacles." He signed off: "Take care, Mike."

So much for hierarchy. Solo wasn't the only one who would be exposed to new training obstacles. I would soon learn to introduce myself as a negative: "I'm Cat Warren. I'm non-LE." Not law enforcement.

Defining myself as not something felt like a huge promotion.

• • •

The city of Durham, North Carolina, for all its virtues, has a relatively high homicide and manslaughter rate—a fact that city leaders avoid in the promotional literature. "Durham. Where great things happen," our city motto proclaims. They do. David and I love Durham. It's safe to live downtown, as we do—especially with an imposing German shepherd and a noisy Irish setter in the house. Enough bad things happen in Durham, mostly just outside my neighborhood, to provide the occasional need for a cadaver dog.

Mike Baker followed through on his promise to expose Solo to new training. "Meet us at the old Liberty Tobacco warehouse at eight P.M.," he said. So what that it hit ninety-five degrees at three P.M.? That's ideal North Carolina dog-training weather. I, on the other hand, have asthma that gets triggered by days on end of code-orange air. I also like to be in bed by ten. Nonetheless, there I was on the dark streets of Durham, searching for the peeling entrance sign to the Liberty Warehouse, walking past the black-and-white Crown Vics with their engines rattling loudly, air conditioners blasting to keep their furry occupants cool. Sharp, intense barks and some paws pounding against the rear door panels made me jump back and then try to correct my wobble.

I had dressed and redressed myself, tucking and untucking my white T-shirt into my Costco cargo pants, trying the shirt outside the pants, tousling my short moussed hair, trying harder than I ever had to look casual: strong but not butch, slightly made up but realistically sweaty. The sweat was no problem. The mousse melted, making my eyes sting. I walked into the warehouse, one of the last standing auction houses for loose-leaf tobacco, once the economic spine of Durham. It was an increasingly leaky and derelict building, now covered with metal siding, and the natural skylights that once let buyers see the color of the brightleaf tobacco were boarded over. The sloping floors that tobacco farmers drove their wagons down were still there, and huge warehouse beams reached up into darkness.

In front of me, I saw a group of men in navy T-shirts and cargo pants standing, arms folded across chests, watching a dog that looked like a shorthaired, sharp-nosed German shepherd with a rattail. I had seen the breed a few times: a Belgian Malinois. I looked more closely at the T-shirts and body type. Two women, thank God. The fawn-colored dog, his tail stiff, was prancing on his toes, circling a pickup truck parked inside the warehouse entrance. He sniffed as he went, tracking along the truck's side and underbelly and around the tire wells. At the back, he froze briefly, nose inches from the license plate. Then he leaped like a cat straight into the pickup bed, wheeled around, and anchored his nose to the spot as though an invisible magnet kept it there, while the rest of his body continued to spin in space. I heard a low gentle chuckle, and someone breathed out slowly: Good dog. Good dog! A tennis ball flipped through the air, landed in front of the anchored nose, and bounced. The game was on. The dog's nails scrabbled for purchase on the concrete as he chased the ball through the disintegrated tobacco leaves and dust.

Mike Baker was still standing with his arms folded. Medium build, medium-brown short hair, medium Irish-English features—the kind of guy who might not stand out in a crowd of noisy dogs and macho cops. But as one handler who comes from another agency to train says, "Mike's magic."

The Durham K9 unit, which boasts between ten and twelve dog-and-handler teams, isn't just for trotting out at city events and school show-and-tells. These dogs and officers work. I finally got used to the background sound of radio crackle; I could mostly ignore it when a patrol car would peel off to a callout: a short track for a burglary or assault, a search for a dropped gun, to sniff a car pulled over on I-85 for suspicion of drugs. Everyone listened intently if a call came in about a child who didn't return home as expected.

Not enough people go missing and are presumed dead for the police department to spend money and time training their dogs to do human-remains detection. It's all about resource allocation—Durham patrol dogs keep plenty busy with all the other calls. It's the same story everywhere: Most law enforcement departments have neither the money nor the pressing need for a dedicated cadaver dog. However, if I could provide a competent cadaver dog, I might fill a gap. Andy Rebmann worked in the salad days of dogs and law enforcement. He started the first cadaver-dog program with the Connecticut State Police. That program has survived up through the present and now has half a dozen dogs and handlers. Programs that spun off from Andy's founder effect—Rhode Island, Massachusetts, New Hampshire, Vermont, Maine—all have law enforcement cadaver dogs. New York and Chicago also have cadaver dogs, as does Brevard County, Florida.

The majority of cadaver-dog work, though, falls to volunteer teams. It's not ideal, from the point of view of most serious law enforcement trainers and handlers. That's not snobbery; it's reality. If you do something for a living, you can spend more hours perfecting and applying your craft. I understood that, but there I was, right in line to try to be one of those not fully respected, jack-of-all-trades, master-of-none volunteers. The good news was that I was in line to learn a lot about dog training. I'd had little exposure to male trainers up to this point—women dominate the obedience world and, though to a lesser degree, the search-and-rescue seminar circuit; men dominate the law enforcement K9 world. That world tends to be a rougher place than the obedience world, where clickers and treats now rule. In the K9 world, square-chain choke collars are ubiquitous and unremarkable, liver treats rare. But woe be unto the handler who doesn't either take off the collar or put it on the dead ring if the dog is going to work, or who takes his temper out on the dog.

Working dogs need the same basic thing, whether from male or female trainers and handlers. Dogs need to hear high happy voices when they're being rewarded. They need to get their reward at the right moment. And they need gaiety.

"Make it fun for him," Mike counseled a too-serious, alto-voiced handler. He modeled the behavior—"Pump him up. Atta boy, atta boy, atta boy!"—pitching his voice ever skyward in an escalating croon.

Solo and I got a chance to train at nine-thirty. The temperatures outside had dropped into the low nineties, but it was about a hundred degrees inside the building. Finally, Mike turned to me and said the even, quiet words that would become so familiar: "Why don't you go get your dog?"

I could barely see Solo's bat-eared silhouette. He had fogged the inside windows of the car I had kept running to keep him cool. I remembered to walk him to let him pee before we entered the warehouse, but that was simply an opportunity for him to realize the threat he faced. Patrol-dog urine coated the weeds up and down the block. Solo walked into the voluminous dark warehouse, stiff-legged and hackled. The scent of all the hyped Malinois and Dutch shepherds permeated the air.

We performed abysmally. I didn't have a search pattern. I was trembling slightly. Solo finally stopped shooting glances over his shoulder to see whether dogs were stalking him and got down to work, moving down the long rows of stacked planks in the rear of the warehouse. He easily found the hides Mike had placed. I saw that from his behavior change. Mike saw it, too. Solo went to each hide, sniffed it, and then walked past it. No alert. Pathetic. I can't remember whether I made excuses. Probably. Mike was sympathetic, complimented Solo on his "work ethic," but explained that he wanted a narcotic-dog alert: nose unerringly pointing to the hide, even if the dog's eyes were rolling up and backward wildly, anticipating the toy reward coming from the handler behind him. Solo, he said, was paying a bit too much attention to "Mom" during the search. That was unusual for Solo; my own fear smell must have been rolling off in waves, distracting him from his dog obsession.

It felt wonderful to put Solo back in the car and return to watch dogs find hidden handlers, fake suspects, perched in the warehouse's massive rafters. The dogs' deep warning barks rang out along with their handlers' standard warnings. "Suspect in the building. This is Durham K9. Come out with your hands up. This is your final warning. Come out with your hands up, or I will send the dog."

The professional term for sending the dog after someone is "suspect apprehension." The informal term is "bite work." Released, the patrol dogs tracked the suspects, found the suspects, and barked harshly. Sometimes the suspect would throw the dog a toy reward, a Kong or tennis ball, and that was the end of the exercise. Other times the suspect would descend from the rafters in a jute bite sleeve, threatening. The dog would launch, mouth gaping, all four feet in the air, slamming into the hard sleeve. The dog would be encouraged to hang on while the suspect struggled, sweat mixing with dust and dog saliva.

"Praise him up," Mike counseled a handler, who ran his hand approvingly over the muzzle of his biting dog, calming him down, getting him to hold hard and securely until he was told to let go, or until the decoy could safely shed the sleeve like a slipped skin, giving the dog the final reward for his work: a big chew toy.

I was fascinated, not horrified, to see so many teeth. It took guts, steady nerves, and perfect timing on the part of all three players: the dog, the handler, the decoy. I stayed well out of the way, high from adrenaline and from inhaling old tobacco dust rich with nicotine.

"Figured you'd have some nerves and that would rattle Solo a bit," Nancy Hook said in her later e-mail to me. "It sounds like you did pretty darn good, though, and did well enough to let Mike know Solo is reliable. I know you can learn a lot from those guys."

Mike was judicious in his critique. Break it down, he said in an e-mail. Worry about one building block at a time. Don't try to control everything. Let Solo initiate the game.

Of course I tried to control everything. I was a border collie in my prior life. Now I had two sources saying the same thing: Let go. I read and reread Mike's and Nancy's e-mails. Instead of having one mentor—Nancy—I had two. I had an entirely different set of dogs to watch. I had a big handful of experienced police K9 handlers and some not-so-experienced K9 handlers to observe. Even with the attendant anxiety, the ambivalence, and my obvious outlier status, it was a deep pleasure to begin again. I could spend part of my time hanging back, studying. I was a student once more, known in the working-dog world as a "green" handler. It would be easy being green. Except when I had to work Solo in front of Mike. Or the other handlers. With their arms folded across their chests. Watching.

While I was at a huge disadvantage, I realized Solo had an advantage here that he didn't have at the local kennel club. Law enforcement handlers don't expect their dogs to get along. Most of their dogs have an edge. Every dog was on lead coming and going; each dog worked separately. The warehouse rang with another warning I would become accustomed to: "Dog in!" or "Dog out!"

For me, that warning was a comfort. A standardization of practice that would benefit me greatly. Working Solo, I wouldn't have to keep my eyes peeled for a shorthaired pointer to come bounding over off lead. Soon enough, Solo realized the same thing: With cops and Crown Vics around, he started to ignore sharp barks and growls and dog-permeated air. I didn't have to apologize for his personality. To the police K9 handlers, Solo wasn't a sociopath. He didn't even qualify as a jackass.

Most important of all, Mike's training philosophy fit Solo's king-of-everything attitude to a T. "Remember," Mike wrote in his e-mail, "we are just anchors holding on to their leash."

. . .

I've interviewed hundreds of suspects that have been confronted with a police service dog and they simply say the same thing over and over again: that is, "Hey, I was willing to fight the police, but I didn't want to fight that dog."

—Terry Fleck, canine legal expert, 2012

I had been invited to watch the last three nights of the twelfth week of training three green dogs and three green handlers in Fort Lauderdale,

Florida. After that, the men and dogs would hit the streets. Ready or not. Steve Sprouse, the trainer for Broward County Sheriff's K9 patrol division, preferred them ready.

Steve said a version of the same thing, with a hint of melancholy, at the end of each exercise: "Remember, guys, it'll be different." Training and actual deployment are separate worlds. He said it to the intense, eager handler who had the equivalent of a furry Mack truck on the end of a leash, to the handler with a dog who needed to develop more spine, and to the handler who seemingly had it all, a balanced dog and a balanced approach to handling.

Sandy and spare, with a slightly drooping mustache, Steve is simultaneously relaxed and wary, ardent and careful. In his late fifties, he's been handling dogs and training handlers for decades. He knew the dry admonitions he sent into the mild late-November night air were falling not on deaf but certainly on naive ears.

Steve and the three green handlers were preparing for the equivalent of a dog-and-pony show: In two days, the teams needed to demonstrate their newly acquired skills before a bunch of superior officers. Steve and the three handlers had discussed how to divvy up the performances. Each dog had his strengths.

One of the three German shepherds had a beautiful obedience routine. Broward Sheriff Deputy Pete Sepot's new dog, Diesel, was great at bark-box work. Pete faced the huge dog toward six human-sized boxes scattered on the training range, and Diesel flowed like sable mercury from one plywood box to the next until he smelled the hidden decoy. The flow stopped, and Diesel balled up at one of the boxes, giving fulsome warning barks, both front feet coming off the ground with the force of certainty. He's here. He's here. He's here. Diesel was just beyond adolescence but already a single muscle from head to tail; his bark sounded like that of a much older dog. It's an important warning: the prelude to more extreme action. The kind of bark that would make many suspects come out with their hands up. That's why patrol dogs are trained to be noisy when they've found a suspect; the bark alert can prevent worse things from happening: Their bite is worse than their bark.

Lughar, another sable shepherd, and his handler, Dave Lopez, had been tagged to show what smooth apprehension work looked like—and as important—how a handler can call his dog off before the dog takes a bite. It's a standard training exercise: If the suspect gives up, or if the handler realizes the dog is headed full tilt toward the wrong person, the handler wants the dog to come back. If it's too late and the dog has already launched at the suspect, at the least, the handler wants to be able to tell the dog "Los!" ("Let go" in Dutch) and have the dog obey him. A dog with real drive has a hard time obeying either of those commands when he's flying toward a suspect; he's fighting his own instincts.

Lughar, though, made everything look easy. He had excelled at turning around and coming back to Dave, then, when he was released once again, going back toward the decoy, getting a solid bite, and coming off the bite on command. He had plenty of drive but was capable of listening. Dave, like Lughar, had everything he needed to succeed—except experience. He was a thoughtful, serious handler.

The problem tonight was that Dave's kids had the flu. They'd been up all night with high fevers. Everyone had been sick. Except the dog. But it always takes two—dog and handler—to make it work. Dave's exhaustion and nerves were running down the leash. "What happened to my perfect Lughar?" Sprouse said mournfully, watching his prize pupil fly down the field, backlit with sodium vapor lights, and dive into the shadows where Pete, playing the role of decoy, stood. It would have been beautiful if Dave hadn't been yelling harshly, in vain, for Lughar to return to his side and ignore Pete. Instead, Lughar ignored Dave: Pete's bite suit was too enticing. Lughar leaped, full-mouthed, at Pete's raised arm. Steve could envision the graduation demonstration, high-ranking officers watching as an open-jawed Lughar kept charging, Dave yelling futilely behind him.

No police K9 function is more misunderstood and more terrifying than the patrol-dog bite. Police K9 units have put a good distance between the civil rights–era protests when Birmingham Commissioner of Public Safety Eugene "Bull" Connor turned fire hoses and snarling German shepherds on peaceful protesters. Warm and fuzzy now prevails when the comfortably middle class thinks about police canines. The media portrayals of police canines these days are "four-footed community police officers," said Charlie Mesloh at Florida Gulf Coast University. A former police K9 handler, Charlie is now a criminology professor who researches the use of force, including the use of patrol dogs.

Although many dog lovers view police K9s as anthropomorphized heroes, catching bad guys, saving their human partners from armed and dangerous suspects, we don't like to think too much about exactly how dogs do that. They don't have opposable thumbs. They have speed. And teeth. Like other uses of force, dog bites can do serious damage. If a handler, or an entire unit, is overzealous, departments can be liable. That's true of any use of force, but a visceral fear arises from the misuse of dogs. Yet the "find and bite" or "bite and hold" patrol dog is the most common patrol dog in the country.

The dog's first job is to intimidate so that a suspect surrenders before a dog or any other weapon is deployed. Terry Fleck, a legal specialist in the use of police K9, is clear: He considers them a "use of force elimination tool."

If the suspect doesn't surrender, the dog's job is to go in, find the person, and bite him. Or, infrequently, her. Unlike guns or even Tasers, police K9s in the United States so rarely kill a suspect that there's only one example in major case law. Just as a comparison, pet dogs and strays do much more damage: They killed thirty-one people in 2011. Tasers caused about five hundred deaths in the last decade. As Charlie Mesloh noted, whenever a particular use-of-force method becomes popular, the incidents of its misuse rise as well.

The occasional misuse of patrol dogs creates enormous bad press, though, which in turn can spur lawsuits and overwrought and under-researched recommendations and edicts that spill over and affect the responsible units. The bad cases have multiple consequences. In 2001, following the media exposure of two K9 units, one on either side of the country, with out-of-control K9 officers, the U.S. Department of Justice decided that it would be better to shift paradigms and recommend that K9 units use a system called "bark and hold" rather than "bite and hold." In other words, the patrol dogs were supposed to simply circle the suspect, or to stand and bark at the suspect until that person tried to use force against the dog or the officer. At that point, the dog was allowed to bite. The DOJ didn't have much evidence on bark-and-hold effectiveness before recommending the change. And thousands of K9 units across the country already trained bite and hold.

Charlie Mesloh decided to look at the issue. After all, the DOJ was recommending a huge change with little research, and three quarters of Florida units used bite and hold.

The concept of bark and hold sounds great. No teeth. The dog scares the suspect and holds him without harm until the officer claps on handcuffs and leads the chagrined suspect away. It's the best of all worlds. The only problem? Charlie Mesloh found, when he did a careful study in Florida, that bark and hold was "a good marketing tool"— and little else. The dogs trained in bark and hold actually bit suspects *more* often than dogs trained in bite and hold. The method inevitably gives the suspect more time to use a weapon on either the handler or the dog. There's no national tracking system for patrol dogs injured or killed by suspects, but the numbers are substantial.

There are lots of good reasons to use dogs for criminal apprehension. If the dog is well trained, the handler can control him in ways he can't control any other weapon. An officer can't recall a bullet once it's left the gun chamber. Once an officer has pushed the trigger on a Taser, he can't change his mind. A dog is different—in principle, at least. That's because good patrol dogs don't arrive with on and off switches already installed. Training a dog to the point where he looks like a well-oiled machine is a beautiful thing, but being able to stop an enthusiastic and driven dog from taking a bite takes skilled training and handling. Lughar's selective deafness wasn't ideal; however, it was understandable and fixable.

. . .

As Steve Sprouse and I drove to the next training area, the aging patrol car's big engine shaking the floorboards, we talked. Not about why law enforcement uses dogs in this manner. In this country, that decision was made decades ago. Ideally, patrol dogs are used to stop danger-ous suspects, and Steve wants them to do a good job of it. The dog should give one solid, full-mouthed bite with no hesitation, no coming off the suspect, or giving the suspect a chance to use a gun or weapon on the dog or officer or bystanders. No more and no less, although that, too, depends on what's happening in a chaotic environment. A dog with a good bite can do less damage than a dog chittering up and down someone's arm, biting and rebiting like a psychotic typewriter. Or nipping, or not biting at all, both of which create more problems. Any patrol dog who hesitates before going in on a decoy during train-ing needs work.

Steve was a green handler himself back in 1989. He had a green dog, a handsome, bold German shepherd named Rick. Rick was a Schutzhund champion. The sport of Schutzhund, like advanced obe-dience in a show ring, is an elaborate and difficult performance that includes bite work.

The call came in one night around eight P.M.: Someone had robbed a convenience store at gunpoint in a strip mall. Steve responded, and soon he and Rick were pounding after the suspect.

Rick went in on the suspect and bit him but didn't hold on. He simply wasn't used to street conditions. Rick had been in Schutzhund competitions where everything was the same: a regulation field, a regulation target, and a regulation bite. This situation was the opposite of regulation. Rick was used to a guy wearing a bite sleeve. The suspect wasn't moving like a Schutzhund decoy, arm pitched at a perfect angle, with a bite sleeve. Rick wasn't stupid; there was nothing to bite. The armed suspect, higher than a kite on drugs, easily got away from Rick's uncertain mouth. Rick figured he'd done enough biting for the night. Steve had to stop the suspect, since Rick wouldn't. By the time backup arrived—and it wasn't long—the officer could see the guy was swinging a gun toward Steve. The backup officer shot the suspect six times before he dropped. In the chaos of struggle, the officer shot Steve as well. The nine-millimeter slug shattered Steve's humerus, severed his radial nerve, destroyed his outer tricep. Steve's arm was hanging by a thread of tissue.

That was when Rick decided it was okay to come back in and "re-engage."

"I could see the dog had the guy by the upper leg. That was good," said Steve in his measured way. "We got success. It was painful success, but we got it."

The suspect lived and Steve lived. But as Steve was lying in his hospital bed, arm sewed back on, he stewed. The backup officer had done what he had to. If Rick had done his job properly, Steve—and possibly the suspect—might not have been shot in the first place.

It would be a year and a half before Steve could go back on patrol. Rick wasn't kicked out of the police K9 business. It wasn't his fault. Instead, Steve, his arm in a cast, retrained Rick. He set up scenarios that mimicked real life, not the Schutzhund ring. He weaned Rick off equipment, so the dog didn't think a bite sleeve needed to be there for him to bite. It's not just Schutzhund-trained dogs who do this; police K9s can all too easily become sleeve-dependent. As one K9 trainer noted of bite sleeves, they should come with a warning label: "Use sparingly for best results."

Steve's cast had a bone-growth stimulator that gave his reconstructed arm regular electromagnetic shocks, a kind of human e-collar reminding Steve of his primary job: Train the dog, train the dog.

Man and dog went back to work in 1990. The first night back on patrol, Steve got a call: a suspect attacking an officer. Steve responded and sent in Rick. Rick bit the suspect. Rick held the suspect. He did his job. And he kept doing his job for a good number of years, until he retired.

Thanks to Rick's initial failure, Steve Sprouse is now a bite specialist, considered among the top aggression trainers in the Southeast and perhaps the country. More important than all the national awards he's garnered, Steve has used his patrol dogs in hundreds of apprehensions without getting shot and without having to use a bone-growth stimulator again.

Steve passes on his hard-earned dog knowledge. He trains his fellow Broward County officers and nearby police units. He also travels across the United States and the world, teaching patrol-dog scenario work, tracking, bite work, and the critical importance of obedience work.

• • •

As Steve Sprouse and I arrived at our final training spot for the night, the automated cyclone fence gate squeaked and whirred. We were at the deserted water treatment plant in Oakland Park, Florida, with its huge stucco buildings and massive wastewater treatment basins. A street light cast bluish light on a strangler fig that embraced one of the fat water treatment basins, inserting hundreds of loopy roots in every crevice. It looked like a vegetative squid, the same species of ficus that pulled apart the stone temples of Angkor Wat in Cambodia.

Instead of Buddhist ruins, we were standing in the shadows of a 1940s-era wastewater treatment plant. Feral cats and raccoons slipped through the shadows. Ferns spilled out of the open ends of cast-iron pipes that once sent treated water back into circulation in Fort Lauderdale. On the tops of the treatment ponds, where paddles once stirred the city's sludge, dirt had settled, creating Jolly Green Giant–sized planters. Forty feet up, it was a jungle. "It's very cool up there," Steve said. In front of us was a huge building, open doors gaping, old equipment everywhere. To our side were a couple of acres of landfill equipment: bulldozers and concrete blocks and massive garbage cans—and more shadows.

Dave Lopez's kids were almost recovered from the flu, and he was less sleep-deprived. Lughar was back to almost perfect, although he was a bit softer on the bite than Steve ultimately wanted him to be.

Steve sent the green handlers and their dogs away from the scene to wait to be called. I offered a hand, and Steve and I went about staging an illusion for the handlers that would teach them to trust their dogs' noses.

The grasses and weeds on the side of the hill next to the plant were thigh-high. The sandy hill was steep, nearly a forty-five-degree angle. As I crawled up in the dark with Steve, almost on my hands and knees, I thought about Florida's coral snakes. I'd never seen one in the wild, but their identifying stripes—"red touches yellow, kill a fellow"— wouldn't be visible in this darkness.

It took us a mere forty-five minutes to set the scene, although it felt longer. We were both sweating in the mild night air. I hauled three big waste containers from the other side of the public works and tugged them up onto the hill, scattering them so they looked like they had been dropped willy-nilly by aliens out of UFOs. Steve found the old hole dug for a training two years before. He cleared it of the roots and weeds that had taken over, and we used our feet to stomp the garbage can down into the hole until its lid was even with the ground. With his penknife, Steve made a small hole in the lid. Casting about the waning beam of a flashlight, I finally found a tough golden graduation tassel on the floor of one of the buildings that would do the trick; we knotted it and ran it through the small hole, so the handler who hid in the can could hold down the lid if a dog found him and decided to dig him out. We tested it and retested it, tugging hard on the tassel, amending it with an extra knot. We pulled brush over the can and stood back and admired our handiwork. The buried garbage can was invisible even a few feet away.

We stashed a decoy handler, helping him into his garbage-can prison. Steve handed him a latex arm. Steve had constructed it from liquid rubber, using his son's arm as a mold. It was a work of art—ductile and creepy—a toothsome reward, just in case a dog and handler managed to work out the elaborate problem. We stood at the bottom of the hill, looking up, bemoaning age and complaining joints.

Steve got on the radio. The first handler, Dave Lopez, arrived in his patrol car, pulling up hard just inside the gate. Steve briefed Dave on what was happening—maybe one, maybe two guys ran this way after a robbery. He warned Dave that feral cats and raccoons were everywhere. That the guys were dangerous. Just over the hill, people were having a late-night barbecue. "Be careful," Steve warned him.

Dave didn't start the search where he should have, given the wind direction out of the northwest. It was a beginner's error, one that gave him a huge advantage. Lughar dragged him around the near side of the mushroom-shaped treatment plant, where Steve and I had retraced our steps, tracking us backward along the bottom of the hill. Then Lughar flipped his whole body. He had hit the decoy's scent coming down the side of the hill in less than a minute.

Steve and I watched with admiration and some dismay as back-to-perfect Lughar made short work of the puzzle. He surged up into the underbrush and pounded the lid of the can with his teeth, muzzle, and feet, making the handler inside howl in misery as his fingers got pinched trying to hold the lid down. There was no time for Steve to play with Dave's mind by telling him it was clearly a raccoon that Lughar was after. That howl was all too human.

It was over in two or three minutes. Steve was discouraged by the fast pace but determined to get some bang for our dramatic stage-setting buck.

Pete and his large sable shepherd, Diesel, were next. Steve was no longer sure that the scenario would slow them down as much as he wanted it to. This time Steve made certain that Pete had pulled well inside the compound, away from Lughar's clever backtrack.

"Hey, Diesel, boy," Steve said. Diesel was already staring out into the dark, ignoring Steve. Steve turned to Pete, frowning. "There are a lot of cats, raccoons, people north of the complex, so just be careful. Give your warning."

Steve watched as the released Diesel set up a solid pattern, moving through the deserted Bobcats and stacks of palettes and concrete culverts on the flat, doing natural switchbacks. "He's got really nice structure." It wasn't clear whether he was talking about Diesel's powerful, sleek looks or his search pattern. Probably both. Diesel worked the area like an old pro, ignoring a cat as it slid away from him, clearing a half-acre with ease. Diesel and Pete were methodical; they were also raising our hopes that they might not solve the problem as quickly as Lughar and Dave had.

"Wouldn't you just love to be inside a dog's head?" Steve asked quietly, arms folded across his spare frame, watching as Diesel swung his huge muzzle, and adjusted his body so it followed his big nose. The breeze had shifted slightly, carrying scent. I could hear the faint sound of Latin music and voices from over the top of the hill.

"There are a lot of people barbecuing on the other side of the fence, so be careful," Steve called to Pete. The admonitions were like whistling in the wind. Diesel, the Mack truck, caught the decoy's scent. He moved uphill in first and then second gear, toward the underbrush and the hidden garbage can. Pete was right behind him, scrambling up the hill.

Steve attempted to distract Pete, but Pete kept following Diesel, who was under the brush next to the hidden garbage can. "So you got somebody or not?" Steve called impatiently. "We need to clear this out now."

Diesel's harsh bark resonated down the hill. He had somebody. He wanted to clear the decoy right out of the garbage can. We could hear his nails scrabbling on the lid.

Steve sighed, shrugged, called to Pete to hook Diesel up, and yelled at the hidden handler to open the garbage can and wave his rubber arm in defeat.

Diesel carried the heavy rubber arm back to the car as a reward, though his enthusiastic growling and head shaking would probably tear a hole in the precious arm, and Steve would have to make a new one. That was the price of rewarding solid aggression work.

Perhaps it was the wind, perhaps the humidity, perhaps beginner's luck. Whatever it was, Steve didn't want cockiness coming out of this exercise. He looked at the three rookie handlers.

"Remember, when you go out into the street, don't think it's going to be just like this. There are three of you, and the dogs have been smelling you for twelve weeks, so they know the odor. They're very confident. You get out on the street, it's a whole different thing. People are yelling, music playing, different odors. You all wear the same style of boots and uniforms. All of that comes into play. It's going to be different."

The handlers were nodding obediently and happily. They clearly respected Steve. I wasn't sure they believed his warning. It was a soft and lovely night. The wind was gentle, the humidity low. It was magical. Steve and I walked back to his patrol car. One final scenario remained before his green dogs and handlers graduated and hit the streets. But only after dinner and a bit of rest.

"In the right circumstances," Steve said with a hint of melancholy, "that's a very difficult find."

The dogs and handlers did beautifully, Steve told me later, at their demonstration at graduation in front of the brass. Lughar's recall was perfect; Diesel was fluid and impressive on the box work. More important, though, is what happens out on the streets. The twelve-week foundation course headed the handlers in the right direction. They will keep training and learning, both on the street and during training. For now, the dogs are doing their jobs. So are the handlers.

Steve was pleased and, as usual, understated. "The dogs are successful. The handlers are the same."

9

Into the Swamp

But there is nothing like work for toning down excessive fear or joy. So I ran ahead, calling him in as gruff a voice as I could command to come on and stop his nonsense, for we had far to go and it would soon be dark.

—*Stickeen: The Story of a Dog,* John Muir, 1909

Patrol cars were parked helter-skelter, radios popping, laptop screens glowing through darkened windows. A few cops, scattered across the apartment project's dismal lawn, looked up as my Camry slid in. I lowered the rear window so Solo could be ID'ed, and he provided his profile, sticking his bear-sized head and chest out the window. One officer hitched his gun belt and jerked his head. Ten-four on the *canis searchus*.

It was a fine afternoon for a search. The morning wind had calmed, and the late-spring air was heavy with unseasonable heat. If scent were out there, the swamp would capture it. It would be hanging like invisible moss on the vegetation, floating above pools of water, waiting for the dog's nose to harvest it.

This was not the first search Solo and I had been on where some central characters were drug dealers. I suspected some were watching us from concrete stoops as I got out of the car and sorted through my minimal equipment in the trunk. The cops were talking to the missing person's girlfriend. I heard sobbing. I leaned farther into the trunk to pull out my backpack, willing myself not to look. It wasn't my business.

This was our first search where climate change probably played an indirect role. It wouldn't be the last. Drought followed by a tempest of rain had caused flash floods. Sheets of water slid off dry, hardened soil into overwhelmed storm drains. Water, silt, and trash cascaded into the already polluted remnants of Triassic Basin wetlands and woodlands.

The suspect's flight reportedly started with a 911 call. The caller told police that a group of men were selling drugs out of a car. Police arrived. They realized one of the men was wanted for a parole violation. When they went to arrest him, the suspect broke away and fled into the woods. Flooding and darkness made the search dangerous. After a patrol dog tried to track him and nearly drowned, the cops pulled out. By then the suspect was long gone, into the woods and swamp beyond. He called his girlfriend from the middle of the swamp that night. That was the last time she heard from him, she told police. His cell phone had gone dead. The girlfriend and his family waited a couple of days before calling the police to say he'd never made it back home.

At the police station an hour before, investigators and I stared at a satellite map of the area. It was acres. Some of it would be muddy; some might be flooded. I tried for calm confidence, though my heart bounced around in my chest.

There in the station, I helped make decisions about how to run the search, trying to sound as though I'd done this a hundred times, even while I confessed my novice status to the participants standing around a table, trying to lower their expectations—and mine. I'd been training with Mike Baker for nearly two years and with Nancy Hook for more than three. It felt like nothing. I slowed down my voice so I didn't sound like an eager amateur. I'm just a volunteer, I said. Dogs aren't perfect, I said. They're just one tool among many. I said all those things, and I believed all those things.

But when cops call in a volunteer, they want results. I knew that as well. A couple of dozen cops were on standby, expecting to make a line sweep of the swamp. Unless Solo and I could find him.

Even though I'm a Yankee, North Carolina woods aren't alien to me. So when the police asked what I needed, I asked for two people to accompany Solo and me. If they were available. Even within urban limits, a few acres of floodplain can contain sinkholes and creeks with eroded sandy banks that may crumble beneath you. Two people could get us out if we got stuck. Those were just the natural hazards. I didn't want to be out there alone meeting the kind of guy we were searching for.

The two investigators who cheerfully volunteered to accompany us arrived at the scene at the same time I did. They were clad for strolling down a linoleum hallway, not slogging through a swamp. The one wearing natty tasseled loafers and creased tailored pants was getting grief from his colleague, who was also wearing office clothes that were only slightly more suitable. The sharp dresser joked about water moccasins and copperheads, but he wasn't laughing that hard. A couple of the bloodhound handlers from the Triad foothills had told me that only people with phobias ever see venomous snakes. I'm not phobic, and I knew that wasn't true.

I wore no loafers with tassels. I had hastily changed out of my cute linen office outfit and looked like an aging Outward Bounder. Nancy would have been proud. I was shod in hiking boots and cross-country-skiing gaiters to discourage ticks carrying rickettsia or Lyme disease; they might even deflect a reptile fang. No hat: I didn't want to look like a complete doofus. A fine-toothed flea comb would do the trick post-search to flick ticks out of my short hair.

As usual, Solo provided extra sound effects—mewling, crooning, sounding more like a great Serengeti cat than a German shepherd. Even his whiskers twitched. He had on his sturdy nylon harness that could double as a handle to suck him out of the swamp if needed. He hated the harness, since it signaled a bit of constraint in what he thought should be an unconstrained life: It meant he might have to work on a long line. Tough titty. It was staying on.

I ran a final mental checklist: water, water bag, bug spray. Tug toys to reward the dog; I hadn't forgotten those as I dashed out of the house. They were as essential as water. We had graduated from one hard rubber Kong to two rope tug toys. One could be thrown straight into his jaws and the other held in escrow. Solo was still a jackass, too possessive of the tug toy, especially after a difficult training. Using two solved the problem. I always had the fun tug in my hand: the one that could get flung far because it wasn't stuck in a stubborn German shepherd's jaws. We no longer had a power struggle at the moment when both he and I should be feeling fine about the work he'd done.

With the permission of the commanding officer, I'd also planted a Mason jar with redolent training dirt a good distance away from the search area. It wasn't a given that we would find anything on the search. Critical to Solo's day ending in a satisfactory way was letting him find some bit of the dead. Mike Baker had given me a gentle lecture after one of Solo's first long searches. We hadn't found anything, and I hadn't brought any training material so that I could give him an honest reward. Solo needed to be paid for his work, just like I did, Mike noted. This time I remembered Solo's paycheck.

My list-checking was simple distraction, a rosary of search equipment that slipped easily through my mind. Solo's distractions were more concrete. Dog piss on the rusted jungle gym and dog piss on the grass, making his tail curl defensively. As we walked toward the woods, children from the apartments—having stood back at first to watch from a distance—now surged around us, chattering like mockingbirds, then leaping back and screaming in exaggerated fear as Solo swung his huge head in their direction and wrinkled his brow. They weren't dogs, they were just little people. He relaxed his tail to wag slow and low, all stiffness gone. Around children, he dialed back both his drive and his rhetoric. "Don't worry, he's sweet," I told them. He wasn't always sweet with me, but he was with others. "He's big, huh? He won't hurt you. He loves kids." He wasn't a therapy dog, though, so we kept moving the whole time.

Before we entered the fickle shade of the mimosas and elderberries and trees of heaven that have colonized and overwhelmed the edges of these Piedmont woods, Solo paused to inhale a last resentful snort of dog piss, baked onto a corroded steel post. Then the two investigators, the dog, and I ducked under the yellow crime tape and through a jagged hole in a chain-link fence.

We straightened up on the other side, and I took one of the rope tug toys out of the training bag and tucked it ostentatiously into my pocket. Solo yowled and spun in delirium, kneecapping me with his big shoulder. I winced and unhooked him from my aching arm. I was tempted to clock him, but I reminded myself once again that searches were not the time for etiquette lessons. I didn't want him to sit or heel or watch me with adoration. I wanted him to find the body.

Unhooked and far more obedient off-lead than on, Solo stood frozen, waiting. His eyes fixed outward, then sliced back to my pocket, then down the hill littered with empty liquor bottles, toddler diapers, a shattered Big Wheel, a rusted washing machine.

He didn't need to hear the command, but the words focused and centered me, reminding me that we'd researched this dance for a couple of years. The least I could do was get it started.

"Solo? Go find your fish."

He did a final brief tarantella around me, striking the pocket where I'd stashed the toy with his open muzzle. Not a bite. It hurt nonetheless. Brat. Then he barked sharply and disappeared into the dense undergrowth.

Solo was bounding, zigging and zagging downhill, the grim apartments left behind, past the distractions of police radios, dog pee, children, and the broken glass and twisted metal that slowed me but not him. We'd breached the first wave of Chinese privet and avoided most although not all of the catbrier, which has the habit of digging its claws deep through thick shepherd coats, right at chest height. We were now in understory, rife with baby buckeye and larger oaks, hairy cords of poison ivy coiling around their trunks.

Then Solo disappeared from view. He had taken his first hot-dog leap, launching off a sandy cliff as it crumbled beneath him. He landed hard in the creek bed below. His maneuver thrilled him. The investigators, distracted from their argument about whether quicksand lay beneath the still-high water, watched with bemusement as Solo tested the two hypotheses. He ran down the stream bed, dropping his jaw to scoop up water and sand like a dredging shovel, humping his back like a porpoise, tucking his tail between his legs, spitting out the sand, then turning around and doing it all over again. It wasn't quicksand.

I explained to the investigators that this particular behavior did not represent Solo at his most disciplined. Purpose and fortitude and focus would arrive momentarily, I promised. I showed no amusement. We were looking for a victim, not playing games.

Solo had already moved on. Having expelled his ya-yas, he put his nose down in the creek bed, then slowed to a crawl and cricked his long expressive tail. He was working. That tail set showed that he wasn't smelling animal, which makes the kink defensive, held higher above his back. Nor was it his human-remains scent crick, where he holds his tail lower and parallel to his back but with an odd piggy curl at the end. No, that was a live human scent tail. He surged out of the creek to the other side, and I could see now where a small portion of the cliff, created by the sandy sediments washed down from roads and construction projects, had collapsed on the far bank of the creek. The sedge grass there was bruised and broken.

I knew the police K9 hadn't made it that far, so neither had the police. Deer, raccoon, beaver, and fox—the only herbivores and mesopredators that manage to survive in these polluted woody swamps— don't create that much of a mess. It must have been a human.

Solo is neither trained nor encouraged to do live trailing, except when I tell him to go find David in the house or yard. Solo's job is to hit the edge of the cone of cadaver scent and then define its parameters, spooling back and forth until he comes down the cone to its tip: ideally, the source of the scent. But in a swamp and woods this size, it could take some time until he hit scent. If he hit scent. So it wasn't stupid to have Solo start his work from where the suspect most likely fled, the point last seen.

Solo didn't care where I started him. He was playing his favorite game. He didn't realize, though, that his favorite game had escalated. He was looking for an entire body, one that probably had been out for four days in a warm, wet environment. Its scent cone might be huge. We were early in our work. Solo hadn't found a whole body before; nor had he been exposed to one, even in training.

If the body were here, it should be easy for Solo to find. He was used to looking for parts per million—a tooth here, a bloody cloth there. While Solo and I had worked on many acres at Nancy's farm, and I had watched him pick up scent from training aids a couple of hundred yards away, I had no way of knowing what he might do in this case. Just a year before this search, Solo trained on a large cardboard box that contained both dirt and a blanket that had lingered beneath a body for some time. The scent overwhelmed him: It's here, no, it's there, no, it's everywhere. He wandered around in the miasma, drunk and giddy. He finally stopped and stood in the middle of the funkiest spot, puzzled. What am I supposed to do with this?

At least one experienced trainer and handler thinks that a version of this dizzy-dog reaction can also happen over great distances and discombobulate the dog and the handler. Deborah Palman, a retired Maine game warden who broke the gender barrier, becoming that state's first female game warden in 1978, looks exactly like what I think a retired game warden should look like: She has close-cropped gray hair, a quiet, competent demeanor, and a sly sense of humor. She doesn't posture. She doesn't brag on herself or her dogs. Yet if I got lost in Maine, I'd give great odds of being found, alive or dead, by Deborah and one of her dogs. Her work has taken her across the vast distances of Maine, from swamps filled with moose, to hills filled with moose, to fields filled with moose—and bear. She even trained fish-detection dogs to keep track of anglers who break fishing regulations.

Deborah and her German shepherds have located more than twenty missing people, some alive, most dead. As a result of working hundreds and hundreds of searches in a wilderness state with varied terrain, and because of working in a team, flanking on searches, reading reports, and training dogs, Deborah has thought a lot about human-remains scent and even live human scent: where that scent can move, how far away, and where the dogs can detect it. How scent can loft over trees and small hills and then come down and get caught or pile up against a hedgerow or an opening "like debris in a stream." Over more than a mile sometimes.

"My young dog, she's very quick to pull the trigger on her indication. She comes into a big pool of scent, she goes wham." Deborah mimics her high-drive female, Quinn, with a squeaky voice: " 'It's here! I haven't really found it, but it's here.' "

There can often be, Deborah said, "a big discontinuity" between where the scent is and where the body is. Where scent disperses can confuse the hunt for both live and dead people. On one case in New Hampshire where Deborah just wouldn't quit, they recorded a dozen alerts from different dogs. Hundreds of volunteers had searched for ten days to cover the territory, receiving scattered reports from people across a mountain valley where they smelled something at various times of day. Deborah's own dog had behaved strangely earlier that day. But Deborah saw the deceased victim only by accident, while she was going up a different trail she hadn't covered in order to find a leash she'd dropped earlier that day. "We're riding on the ATV to get into town, and being a game warden, I can't help looking into the woods, and I spot her body off in the woods down the trail. Finding her was dumb luck."

The victim was lying just off a walking trail in a shady glen. Deborah said it was an enormous lesson for her—once she'd gone back and studied maps and reports and thought about her dog's behavior over the days of searching, and even that of the other search dogs. "Your dog's going to go crazy, looking and looking and looking and not finding it," she warned handlers at a seminar.

So you have to know how to read your dog. And the wind. And the temperatures over days. And the water and terrain. Everything, really. Finally, it can come down to something like chance. Like dropping a leash.

Deborah paused. "So, who made me put the leash there?"

. . .

In North Carolina, Solo, I, and the investigators were past the wood line and down in the swamp. As far as I could see was waist-high neon-green poison ivy. Hundreds and thousands of immature *Toxicodendron radicans,* with leaves the size of a toddler's hand, waving in the breeze. Growing together closely enough that I knew I couldn't weave among those open palms without their touching me. Solo was demonstrating the inevitability of getting tagged. I could barely see him, about a hundred feet away, as he created a lively conga line of poison ivy. He plowed through, harvesting the oils so they would rub off on me next time I touched him. Over the last decade, poison ivy in this region has been getting bigger, growing faster, and becoming more poisonous.

Poison ivy loves global warming. Here, in the open swamp, it didn't have the opportunity to train itself into the hairy ropes one sees on trees, but there was something more disturbing about these tender ivy infants. I had to remind myself that, as noxious as I find it, it's a native plant that feeds the locals: Songbirds from the catbird to the Carolina wren, honeybees, deer, and muskrat benefit from its tiny blooms, leaves, and berries.

That sea of poison ivy could also be hiding a body: someone who had run from the police at night in a downpour without a flashlight. He wouldn't have thought about poison ivy; the dark and rain would have hidden its identity. His body could be out there in the middle of it. It wasn't convenient, but I didn't have much choice: I pulled down my long sleeves, buttoned them, raised my arms above the ivy, and followed the leader. The two investigators followed us through the bog. Their loafers were getting wet, but I heard no complaints.

We had cleared a section of the swamp when I suggested a temporary halt to cool the pooch. Solo had quartered back and forth gamely, no longer on a trail, simply trying to catch the edge of a scent cone as he'd been trained. *Nada*. On the bright side, none of us had drowned or hurt ourselves. We'd been out for only twenty minutes or so, but it was eighty degrees. Solo, dashing around in his double-fur coat, was not quite hyperventilating but close to it. He was in great physical condition; nonetheless, a scenting or tracking dog on the job can tire much more quickly than a dog out for a walk—a scenting dog isn't just breathing, but is deliberately pulling more air in, and sending that air in a different direction once it's in his nose, to identify the scent. A sniffing dog breathes in between 140 to 200 times a minute, compared to a dog out for a stroll, breathing at thirty times a minute. Solo, being Solo, was running and sniffing.

We weren't lost, but a certain aimlessness hovered on the edges. Time to regroup. Solo found a patch of muddy swamp and flopped, glassy-eyed, head flung up, dark lips pulled back to capture more oxygen. I poured him fresh water from my backpack—no need for him to get a mega-dose of oil, chemicals, and pesticides from the storm water dumped during the last rains.

We three humans reoriented ourselves, eyeing the map, squinting, pointing to various landmarks on it, finding the corresponding creek beds and electrical towers around us, figuring out the closest street, which way the suspect might have fled. Our conversation assumed a certain rationality on the part of the suspect, and that wasn't a given. On the other hand, if he knew the area at all, it was likely that he'd moved in the direction he told his girlfriend he thought he was moving. The water had receded greatly; I could see higher-water mud markings all around. Nonetheless, it was difficult to know exactly how high it had been that night or even where it had flooded.

I realized we should have started the search from the opposite side of the swamp. Solo might have been following a human trail at the beginning of the search, but there was a distinct possibility that we had been searching upwind of the victim. That wasn't good. We should have started the search downwind. My temporary paramilitary obedience to my police escorts had undermined what good sense I had. With that sense of dread, my feeling of being a know-nothing newbie reemerged. But soldiering on at this point was best. I picked up Solo's water bowl, flipped out the remaining ropes of saliva and water, and hitched the daypack onto my sweaty back.

It felt longer but was probably only five minutes later when I saw Solo slow from his steady lope. He lifted his head. He had started hitting positives as he sampled the air. He threw his head higher, to gather in more news from the rafts of air. We were coming out of the swamp, approaching a copse of trees and heavy brush. Solo angled toward it and slowed even more. He lifted both front feet off the ground in a rearing motion, bearlike, almost bipedal. He looked as though he were trying to climb an invisible mountain or break free of a lead keeping him anchored. He approached the edge of the trees, the sycamores, elm, and sweet gum. With his tail tensed into a tight curl, he moved toward a couple of trunks and peered with suspicion up into the branches. I knew that curl. Solo was in cadaver scent.

Although I suspected this particular victim was not in a tree, I felt a flash of pride about Solo's skepticism. Dogs far too frequently search the ground obsessively, as though it is the source of all scent. One of our standard trainings involves hanging material in trees and bushes, forcing Solo's nose up. Dead people are found in trees more frequently than one might imagine, in suicides and homicides, in massive floods. Even in death, the world is three-dimensional, not just a flat plane.

I felt a bit dizzy. I knew it was my adrenaline surging. This wasn't training, and what Solo was smelling wasn't swamp gas. The cops were well behind us. I wasn't sure where. I'd been too intent on Solo.

Solo kept following his nose, and I kept following Solo. We were now into the shade of the trees on the far side of the swamp. As my eyes adjusted, I saw that Solo had stopped. He was just standing there. A body lay in the shady copse in front of him. The victim was thirty feet from me, less than ten feet from the dog, simultaneously ashen, because of the dried mud, and dark. Facedown, shirtless, mired in mud. I smelled nothing. But the trees and bushes had helped concentrate scent for Solo.

Solo looked back at me. What next?

I fumbled as I dragged the tug toy out of my pocket, feeling it catch and then give. I let my voice escalate, making it merrier with each enunciation. Good boy. Good boy. Good boy! What a good boy! Good fish! Yeah!

Solo was happy to come toward me and get his reward as I backed up. Then he and I were swinging together like two planets spinning in a mud orbit, held together by the rope of a tug toy. I pulled him farther and farther away from the body trapped in the silt. I called over my shoulder, as loudly and matter-of-factly as I could, "Found him."

Over Solo's happy growls, I could hear the investigators' faint, surprised voices. They sounded far away, but one of them repeated himself so I heard his mantra more clearly as they ran toward us. "You're kidding. You're kidding. You're kidding."

Solo's work was done. I let him keep the tug toy. He had earned it. I hitched him up, and we turned and walked away from the canopy of trees, back into the sun and the iridescent green of the marsh.

Cleverness and Credulity

Most people are rather confident of their ability to recognize bullshit and to avoid being taken in by it.

—Harry G. Frankfurt, *On Bullshit,* 2005

Clever Hans, a smart German draft horse at the turn of the twentieth century, drew crowds with his amazing responses to numerical problems. He would paw the ground to enumerate sums and differences. His owner, a math teacher with a white beard that flowed as munificently as his horse's tail, would ask him, "If the eighth day of the month comes on a Tuesday, what is the date of the following Friday?"

"Der Kluge Hans" would tap his hoof, answering correctly, and the audience would burst into amazed applause.

We know why Hans could do this, partly thanks to German psychologist Oskar Pfungst's lengthy report in 1907: He was stomping his foot until he got a subtle, albeit often unconscious, cue from his trainer or his audience. Pfungst's report went further, though. The folie à trois had escalated over time; the feedback loop between horse, handler, and crowd grew with Hans's fame. So did his owner's increasingly delusional faith in the horse. When Hans erred in his foot-stomping answers, his handler stopped noticing.

Pfungst wrote, "One day it came to pass that the horse even understood French, and the old gentleman, whose apostolic exterior had always exerted a high degree of suggestion upon his admirers, in turn fell captive to the spell of retroactive mass-suggestion. He no longer was uneasy concerning the most glaring kinds of failure."

When I first read about Clever Hans and his deluded owner and audience, I viewed the whole affair as a turn-of-the-last-century anachronism. But Clever Hans returned to haunt me during Solo's training. Early on, I thought the story was about the horse, not the human, and about Solo, not me. Now I know better. People use animals for a variety of faith-based practices. Once you start to load expectations on a horse's back, you realize it can hold all sorts of ideologies and theories. The Darwinians saw in Hans clear evidence of the similarity between the human and the animal mind. Cartesians argued that Hans was a mere brute. Hans, of course, was neither genius nor brute, but a smart and devoted horse.

Solo is clever and devoted as well. Which means he's fully capable of a dog version of lying. It's my job as his handler to prevent him from doing that. I don't always perform perfectly. Here is one example from Solo's training, but it's not the only one.

We were in an abandoned warehouse in Durham several years ago. A patrol-dog handler had obligingly put out some training hides for Solo. They had "cooked," as one says about all sorts of scent training material (not just cadaver), for about a half hour. Long enough, on that warm North Carolina night, to send out scent, which I then sent Solo to find. The other handler was idling behind the two of us, monitoring us but not closely. Solo hurtled through the building, accelerating madly, trying to dig his claws into the slick concrete for extra purchase.

His head twisted, he flipped almost in midair, flew back toward a garbage can, and came in for a deep sniff, working scent hard. Classic. I caught up with him and slowed to a stop, admiring his technique. Solo eyed me, and I stood there, stupidly meeting his gaze. He then went into his down alert, staring at me happily. It felt wrong. I looked back at the other handler, who had tuned in a few seconds too late and gave a quick head shake. Nope. No cadaver hide there.

My bad timing, when I slowed down and stared at Solo, helped trigger a false alert. I unconsciously encouraged him to do what he shouldn't have done. Even a microsecond's hesitation makes a difference at certain stages of training. Solo's behavior could become chronic if encouraged. I had made a sloppy beginner's error. The only mistake I didn't make at the garbage can was rewarding him for his minor perfidy. I broke his gaze and repeated the command: "Go find your fish." My voice probably had an edge, although it shouldn't have. Solo moved on to find a couple of hides that were really there.

"That's why we call it training," Mike Baker said after I had put Solo in the car, chagrined. He was slightly irked at me for allowing Solo to con me and at the other handler for not preventing the con.

The adaptive advantages of our two species co-evolving are obvious. Yet those same connections become a real disadvantage when you want a dog's nose to be an independent and disinterested witness.

False alerts have to be dealt with honestly, or they become the elephant in the training room or on search scenes. Like alcoholism, they are more common than acknowledged. A few handlers swear that their dogs never, ever false-alert and that something must have been there, even if it was just residual scent. That comforting circular story—*there must have been something there, because I trust my dog, who is perfect and would never, ever misrepresent*—can come back to bite you in the butt. Philosopher Harry Frankfurt calls it "bullshit." Bullshit, Frankfurt notes in his famous essay, is more insidious than a lie. While someone may not realize she's bullshitting, and while the bullshit isn't always false, it creates a broader problem: a general indifference to facts. We see this in politics all the time—and in the dog world, which is different but similar. That's why Frankfurt argues that bullshitting is more corrosive than lying. The liar, unlike the bullshitter, is aware that he's positioning himself against what he thinks is the truth. Lying takes a bit of effort, a slight respect for the truth that is out there, somewhere.

I will never know for certain, when Solo false alerts, whether he distinguishes between lying and bullshitting. I believe Solo tries to be honest. Mike Baker once called him one of the most honest dogs he knows, partly because it is so easy to read his body language before an alert.

Nonetheless, Solo will false alert. Infrequently, but it goes into my training and search records. Every alert gets counted. It mostly happens when he's in scent, but not as close as he could and should get, but he decides that's good enough for him. Or if I'm handling him badly, as I did in the warehouse. Sometimes I'll never know why. Sometimes I can speculate why he alerted. Not all of them are false. If we're searching a junkyard of wrecked cars, and Solo alerts on a front seat where an airbag has deployed or the windshield is shattered? I'll reward him for that one, even if there's no body in the trunk. Blood can stick around for years.

If there are five or six cops standing and staring at something redolent lying on the ground, say, a bag with a dead dog in it? Solo will look at it, too, look around, gauge everyone's expression, and think, Hey, maybe that's something worth alerting on. They're into it, right? In that case, no reward. Move on. During searches, if people want me to check garbage bags that look suspicious, or particular bones, I politely ask if they can keep a short distance away. On a recent search, I was having Solo check what I was sure was a pile of sand deposited by a recent creek flood, though one careful investigator thought it looked gravelike. I simply asked the searchers to stand away while I ran Solo through the area. They all obediently stepped well away and turned their backs, but they couldn't help looking over their shoulders to see what he would do. Solo sniffed and moved on. Over the years, he has been increasingly proofed off dead animals. More important, he's been increasingly proofed off the human gaze.

It's hardly dogs' fault: We humans hardwired them through selective breeding to be utterly responsive to us. With working dogs, we take it a step further. We ask them to be both deeply bonded with their handlers and to act independently. They need to be both obedient and to think for themselves. We train them to ignore us and go get their work done. Push that door. Don't look at me to do it. Open that gate. Find that body. Stop watching me and do your job. The game is to be together *and* separate. Bonded *and* independent. For some breeds, and for some dogs, it's easier than for others.

It's partly why one of the hardest exercises in advanced obedience can be the "go out." An obedience dog is used to being fed treats in exchange for gazing lovingly, unstintingly into the handler's eyes on the heel and recall. Then the handler asks the dog to run enthusiastically straight away. If it's not taught with the proper chain of treats and rewards, you can witness an otherwise fine obedience dog walk away slowly, sulkily, gazing back at his owner: You don't love me anymore. You want me to go away.

We humans are hardwired, too. We are attached to our dogs. The handler or even the helpful trainer can unconsciously play an unhelpful role in creating dependency, wanting the dog's success. That's why a handler training a sniffer dog should start training on blind problems, where the handler doesn't know the location of the hide and can't help the dog cheat. Then she should go on to double-blind problems, where the trainer flanking her doesn't know the hide's location, either. That's why it was good when Nancy Hook started forgetting where she put the hides in the fields and woods early in Solo's training. She was un-consciously providing all three of us with double-blind trainings. She was helping us avoid bullshit.

. . .

So what's the harm with a cadaver handler here and there saying, with some bravado among friends at a seminar, that her dog never false alerts? Or a bloodhound handler bragging that his dog can follow a two-month-old track, or trail someone driving in a car for miles? Aren't they the harmless equivalent of big fish tales?

No. Bragging on your dog provides a tiny contribution to the general spreading of bullshit about working dogs. It creates a wish-ful blindness that doesn't just end up hurting the training of a par-ticular dog. It also helps create a filmy fiction about working dogs in general—a kind of milky, soft-focus portrait that helps us practice Hero Dog Worship.

False claims, repeated often enough—as tracking trainer Tracy Bowling pointed out—reach the level of legitimacy. From there, one can trace the real and obvious harm those lies create. They undermine truthful handlers who don't overreach. They keep people from training their dogs to the necessary level. They can make the work of a good de-pendable dog look obvious and simple when, in reality, it's enormously difficult.

The exaggerations send the media into a tailspin about the wonder of dogs, then a counter-tailspin when the inevitable cautionary tales emerge. The bed-bug backlash is a nice example. The *New York Times's* honeymoon with bed-bug-detector dogs ended in less than a year. Its first March 2010 article had nary a doubt about the effectiveness of canine versus bug: "Bedbug-sniffing dogs, adorable yet stunningly accurate—entomology researchers at the University of Florida report that well-trained dogs can detect a single live bug or egg with 96 per-cent accuracy—are the new and furry front line in an escalating and confounding domestic war." Dogs in those accounts seemed to work alone, without handlers. They just took a taxi to inspect Upper East Side hotels by themselves.

Eight months later, the *Times*'s tone had changed: "Doubts Rise on Bedbug-Sniffing Dogs": "But as the number of reported infestations rises and the demand for the dogs soars, complaints from people who say dogs have inaccurately detected bedbugs are also climbing."

Cautionary scientific studies are starting to appear, to the great consternation and sometimes rancorous objections of some dog handlers and organizations, who can pick holes in a study faster than an army of termites and claim that they are "shocked, shocked" to find there is anything amiss in the magical kingdom of working dogs. The feigned shock is another nice example of what Frankfurt calls bullshit. But it's inevitable and, at one level, understandable. The canine legal arena has become incredibly complex and contentious—the Supreme Court took on its first two Florida dog-sniff cases in 2013. The cases were based on the Fourth Amendment's protection against unreasonable searches. While the court sided with the dog and handler during a traffic stop in one case, the other case, which involved suspected marijuana growing in a private home, had a different outcome. The police used a drug dog's sniff at the door of the house to establish probable cause to get a search warrant. The majority ruled that K9 noses aren't that different from prying government eyes. Your right to privacy extends to keeping a K9 nose away from your home. That decision will have repercussions on how sniffer dogs' noses can be used.

So when a scientific study casts any doubt on the invincibility of working dogs, handlers and trainers react with alacrity. That was certainly the case for a 2011 study in *Animal Cognition* by researcher and former detection-dog handler Lisa Lit and two colleagues at the University of California, Davis. The study showed that law enforcement K9 handlers, when they expect to find gunpowder or marijuana in a certain place, either will state or will actually believe that their dogs have found the substance—even if there's nothing there.

It was a simple study at one level: no drugs or explosives planted anywhere. Instead, the researchers placed little pieces of red construction paper and told the handlers that those markers indicated either drugs or explosives. When handlers saw a red piece of paper falsely marking a scent location, they were much more likely to say that their dogs had signaled an alert. Conscious? Unconscious? Perhaps both. What was interesting was that the handlers were more distracted by those misleading pieces of red paper than their dogs were by the Slim Jims and tennis balls stashed in various corners. Lit and her colleagues recorded literally hundreds of false alerts.

Lit's study was not a dog study; it was a human-nature-with-dogs-added study, and it highlighted the need for a strengthened training regime. It also pointed to the problem of expectations. If we go in expecting to find something, the chances are higher that we will. We all have confirmation bias. How much better it is when we can get our dogs to confirm that bias as well.

After reading Lit's study, I began incorporating more negative searches in my training with Solo. The first few times I did it—searching an entire abandoned aviation building without a single hide—Solo yowled in protest, trying to get at the tug toy in my pocket on the way out. He was mad. Here he was, surrounded by sheriff K9s and guys in uniforms who love to play tug. And no hides? My pocket got stained with resentful saliva on the way out, my thigh a bit bruised. He didn't false alert. It was a start.

The next time, the negative search would have to be blind: I wouldn't know there weren't hides out. Then double blind, where the person with me wouldn't know if there were hides or not. At some point, I might graduate to the equivalent of pieces of red construction paper and not react in a knee-jerk fashion. One step at a time. I did call home as we left the training area. David put a cadaver hide out in the yard, so as Solo and I walked from the car to the house, Solo's head flipped. He ran toward the scent. Look! Cadaver after all! Give me my toy now. He was pleased. I was pleased that I hadn't needed to give a specific command for him to find the hide. Mike Baker had told me at the beginning of training that Solo should be prepared at any time to define the game without waiting for my specific command.

The study by Lisa Lit and her colleagues is not the only detection study that has shown less-than-stellar results and pointed to the need for strengthened training regimes. Larry Myers of Auburn University did an extensive, not-yet-published study of twelve dog-and-handler teams who work full time at detection.

"It was a simple test," Myers said in his straightforward way. "I was afraid it was going to be too simple." He randomly placed scent samples in brand-new pizza boxes. He stayed out of the room where the boxes were placed so he wouldn't unconsciously cue the handlers. The reliability of the dog-and-handler teams ranged from one team's dismal score of 30 percent reliability to another handler who had 97 percent reliability. That highest score belonged to an experienced trainer and handler who, Myers noted, works double-blind problems constantly.

A big clump of teams came in between 60 and 85 percent reliability. Eighty-five percent is quite respectable. Sixty percent, not so much. That starts to get closer to chance. Thirty percent reliability should make you think about changing the dog or the handler or the entire training regimen.

"It was interesting to see how bloody awfully a lot of people did who thought they were hot shit," Myers commented. "I have given up being amazed at how people can think they're honestly doing something right, and how self-deluding they can be."

. . .

He has testified under oath, for example, that even though he does not keep detailed records of his activities he knows that his dogs have almost never been wrong. According to [Keith] Pikett, as of 2009 his dog "Clue" had been wrong once out of 1,659 lineups. "James Bond" had been wrong once out of 2,266 times. "Quincy" had only been proven wrong three times in 2,831 lineups.

—Innocence Project of Texas report, 2009

Although it's rare, extreme canine versions of Clever Hans have appeared in America's courtrooms, with verdicts of guilt or innocence at stake. Math tricks played for the amusement of crowds can morph into dog tricks played in front of gullible juries, sending innocent people to prison. When handlers lie about or exaggerate their dogs' capabilities under oath, it poisons the well for handler testimony and the credibility of the dog's nose.

This is the kind of testimony that exercises Roger Titus, vice president of the National Police Bloodhound Association. Over the past decades, he has worked trails with his many bloodhounds that made him proud. That work has helped put guilty people in prison. When his dogs are able to follow three- and four-day-old trails, he's incredibly pleased. What undermines the work are the lies he hears in training and on the witness stand. The stories can become albatrosses around the necks of conscientious trainers and handlers. "On occasion, it has become outrageous," Roger said of handlers' claims. "Four months old? Impossible. People who put trails out in January to run in May are full of it."

The danger signals are clear, Roger said. "It's the handler who wants to be a legend in his own mind." Yet such legends end up as sworn evidence in the courtroom and cautionary tales in law journals. Scent evidence, or a dog's sniff, should be one piece of many pieces of evidence in a case, but sometimes it's the major evidence. That's a problem.

Now-deceased Pennsylvania State Trooper John Preston was one such legend. His fraudulent claims of his dog's ability to track scent led to as many as sixty people being convicted solely or partially on his false testimony, according to the Innocence Project. Preston claimed what tracking-dog experts say is impossible—that his dog could smell human traces months or even years after a suspect walked over the ground or on heavily trafficked streets. One man was freed in 2009 after spending twenty-six years in prison. Florida prosecutors hadn't bothered reviewing Preston's cases after he was exposed as a fraud in the mid-1980s. In 2008, Florida State Attorney Norman Wolfinger ordered a review of murder and sexual battery cases where Preston testified, although the local newspaper editorialized that an independent investigation was needed. That didn't happen. Preston died in 2008.

Keith Pikett, a now-retired Fort Bend County, Texas, sheriff's deputy, is a more recent, still-living legend. His claims about his bloodhounds' scenting abilities resulted in what the Texas Innocence Project told the *New York Times* amounts to fifteen to twenty people in prison "based on virtually nothing but Pikett's testimony."

Pikett had been involved in helping indict more than 1,000 suspects nationwide. His specialty was the scent lineup. A scent lineup starts with collecting scent from a crime scene, then collecting scent from a suspect. The dog's job is to "match" the scent from the crime scene with the scent of the suspect. For scent matching to be valid, it needs to be done under pristine circumstances, double-blind, with careful preservation methods. In the Netherlands—where the courts accept scent lineup but only as corroborating evidence—they use more than one dog, and the work is done in a sterile room without handlers present. In other words, no cross-contamination and no possibility of Clever Hans. That's not the way Keith Pikett did it in Texas.

Ultimately, the police evidence videos showing Pikett and his dogs running scent lineups "cooked him," Roger said succinctly. I watched them online. Paint cans with numbers were placed on the grass in a line. An investigator pulled gauze pads in plastic bags out of one can and put them in another with bare hands. If there ever were an uncontaminated scent object with the suspect's scent on it, that scent was now possibly in several cans. Pikett then ran his bloodhounds on leashes down the line of paint cans. The dogs would look up, bay, stop when Pikett stopped. They would shake their heads, slobber flying, and bay again. They avoided some of the cans. Pikett stopped one dog at a can with the leash, and the dog stood there. Another dog paused squarely between two cans, and Pikett said the dog had alerted on one of them. One dog bayed and ran past two cans, and Pikett said the dog had alerted on one of them. Head shakes, barks, and pauses were all alerts, according to Pikett. The bloodhounds were doing all three of those things.

"This is the most primitive evidential police procedure I have ever witnessed," Robert Coote, the former head of a British K9 police unit, testified after he watched the videos. "If it was not for the fact that it is a serious matter, I could have been watching a comedy."

The problem is that cops and prosecutors and juries across Texas bought the comedy for years. One man accused of killing three people, based greatly on Pikett's dog evidence, was partially blind, handicapped with diabetes and bone spurs, and physically incapable of committing the murders Pikett linked him to. He spent seven months in prison before someone else confessed to the killings.

Michael Buchanek, a retired sheriff's department captain, was identified by Pikett's dogs as the prime suspect in the rape and murder of his next-door neighbor, a social worker, based on a police theory that Buchanek had put the body in his car trunk, driven five miles, and dumped the body in a field. Pikett's dogs supposedly followed the victim's scent in a moving car for five miles, twenty-four hours after the crime occurred. As international working-dog experts Resi Gerritsen and Ruud Haak noted with heavy irony in their book, *K9 Fraud,* it was "an exceptional performance that no dog can copy."

The police, Buchanek told the *New York Times,* "just kept telling me, 'the dogs don't lie—we know you did it.'" After months of living under a cloud of suspicion, Buchanek was cleared when DNA implicated another man who later confessed to the crime.

Juries are especially vulnerable to dog testimony, Roger Titus said. "You see them look at each other," he said. "Out of ten people, you've got eight that like dogs. A receptive audience."

Coote and Roger were not the only ones horrified. Roger's colleague Doug Lowry, the president of the National Police Bloodhound Association, testified against Pikett, saying he was doing "a disservice to police bloodhound teams throughout the country." It's rare for organizations or top handlers or trainers to testify against other handlers. But these men believed that Pikett and his practices needed to be stopped. "Pikett has done a lot of damage to the veracity of dogs in the Texas system," Andy Rebmann said.

While Pikett is retired and no longer testifying, his cases still pop up in the news. In 2007, Megan Winfrey of East Texas was sentenced to life in prison for a murder she was charged with committing at the age of sixteen. The major evidence against her? Keith Pikett's scent lineups. On appeal, her father was exonerated for the same murder. Her brother was tried for the murder as well, but his attorney argued strenuously against the scientific validity of Pikett's scent lineups; her brother's jury deliberated thirteen minutes before finding him not guilty. Megan Winfrey appealed to the Texas Court of Criminal Appeals in April 2012. On February 27, 2013, she was acquitted of all the charges against her, but the prosecutor in her case requested a rehearing. She was finally released on April 19.

Pikett's attorney told the *New York Times* in 2009 that his client's work with his dogs could seem mysterious. "The first time I saw it, I couldn't understand what the dogs were doing." But, he added, Pikett clearly knew. "He's been doing it so long, he doesn't understand why we don't see it."

. . .

Corruption, wrongdoing, and cheating exist across the human spectrum. People are smart, just like dogs, so they sometimes cut corners to get their reward more quickly. The vast majority of experiments in cheating show that most people, given the choice and opportunity, will cheat a little. (Most people don't cheat a ton, because cheating a lot makes us feel too guilty. Unless we're Bernie Madoff.) While we are reasonably tolerant about small levels of cheating, when people use dogs like puppets to create a sideshow, we feel especially duped and betrayed. Those cases end up getting an inordinate amount of attention.

Every sniffer-dog and trailing-dog genre seems to have a handler who becomes emblematic of that dishonesty. And every one of those handlers was enabled by people who should have been suspicious. In the case of cadaver dogs, federal agencies, prosecutors, law enforcement, and even archaeologists contributed to the corruption. From the beginning of my training with Solo, one name kept getting dropped with an occasional covert glance at me. Sandra M. Anderson of Midland, Michigan. A volunteer cadaver-dog handler. Like me.

People would ask during training or even searches if I had heard of her. Yes, I had. Almost every cadaver-dog handler has heard of her. Like Pikett, she has harmed the reputation of everyone who works with dogs' noses. Like Pikett, she is a fine cautionary tale.

Anderson started with—as a search-and-rescue handler who knew her told me—"a really good dog," a Doberman-pointer mix named Eagle. Her dog found people. But at some point, Anderson, craving more attention, started to plant bones at crime scenes and at mass graves. Subsequent findings indicated that she was planting false evidence as early as 1999.

Like Keith Pikett or anyone who gets away with doing something more than a few times, Anderson had enablers, including the FBI, who thought she was wonderful. Gullible law enforcement investigators and archaeologists called her dog's abilities "mystifying" and "eerie." That language alone should have been a red flag.

FBI agents arrested Anderson in April 2002 during a search in the Huron National Forest in northeastern Michigan. Michigan resident Cherita Thomas had disappeared more than two decades before, and police continued the search for her remains. Anderson offered assistance. She was arrested after a crime scene investigator and a cop witnessed her planting bone fragments and bloody carpet fibers in and around a tree stump and in the muck of a drained forest creek.

The FBI ultimately had to review hundreds of cases that Anderson had worked on in Ohio, Indiana, Wisconsin, Louisiana, Michigan, and Panama. The early credulity of everyone from the FBI to anthropologists was matched by the angry pendulum of backlash. In her guilty plea, Anderson admitted that she had planted a bloody saw, a toe, carpet fibers, and bone.

The net effect of Anderson's acts was both complex and simple: It made law enforcement even more suspicious of volunteers. That's understandable but not entirely rational. Healthy skepticism should reign in all work that involves prosecuting someone for a crime. Dogs' noses should be just one of many tools. They can be great, and they can do things other tools can't. They aren't mystical, and they aren't perfect. They do need to be reliable.

. . .

There's no shortcut to reliability except constant diligent training. After I had put Solo back in the car that discouraging night in the warehouse, I joined Mike Baker at the other end of the hot, dusty building, where he was working with a green handler learning how to "detail," running his hand up and down near the storage shelves, sketching W patterns in the dusty air, asking the dog to "check up," "check down," "check here." The little Malinois was already panting, hyperventilating. At a certain point, the tired dog paused and started to lock in on a box with his nose, but Mike murmured behind the sweaty handler, "Keep him moving." A false alert averted.

Timing is important at any stage, but it's essential early on. I was learning not to slow and hover but not to move so fast that Solo overlooked something important.

If the drugs or the gunpowder or the bone is actually there and a handler tries to move on? The dog learns how to "commit," to plant himself stubbornly and ignore the handler's prevarications or even a slight jerk on the lead to come off the scent, a pull that a less-evolved working dog might respond to.

It's not mystifying. It's not eerie. It is a beautiful sight, a dog trusting his nose, ignoring his handler's efforts to get him to unstick himself from the flypaper scent that he's stuck to. The dog who ignores the handler's gaze, which is irrelevant to the task at hand. This is what real faith should look like—hard and unwavering. This is what the co-evolution of a working dog and handler should look like. The dog's commitment to the truth in the face of your moving away. That's real teamwork—the dog pointing his nose or paw or entire body at the scent, telling his handler, You bloody idiot! It's here!

11

All the World's a Scenario

This wide and universal theatre
Presents more woeful pageants than the scene
Wherein we play in.

— Duke Senior, *As You Like It,* Act 2, Scene 7, The Forest

Roy Ferguson, a tall hound of a Tennessean, arrived at the dress rehearsal decked out in a fluorescent orange sweatshirt and a tan vest covered with flaps and pockets, gadgets and badges. He looked like an ideal Boy Scout troop leader: geeky and capable of goofy humor, yet stern enough to keep high jinks at bay, and with a handy tool to fix any problem. He stood inside the wood line on a misty winter morning, briefing a laconic dog handler from Florida. A young pit bull terrier, dark brindle with ghostly amber eyes, stood at the handler's side, her sheriff K9 vest providing the only bulk on her slender body.

We were seventy-five yards down a gravel logging road in Georgia that wound into the woods and disappeared. The kind of road that someone might drive down at night when he's panicked and looking for a place away from traffic and homes to dispose of a body.

An eighteen-month-old toddler was missing, Roy told Benjamen Ortiz, the handler. The child was reportedly dismembered and buried. Law enforcement had discovered a possible grave site down the trail behind them. "Work your dog in the area, come out, and tell me what you've found."

Ben nodded and released June Bug. She bounced down the trail like a gazelle, beelined to a mound of freshly dug humus, sniffed long and carefully, went to another mound, eyeballed Ben, and snatched something off the ground. Good reflexes are critical in this work.

"Get back to work," Ben snarled softly. June Bug skittered sideways and levitated over a log, still munching the mysterious delicacy. It gave her enough sustenance to settle down to work. Ben stood back and watched without speaking. She moved on and so did he, quartering back into the woods.

Five minutes later, though it always seems longer, Ben reported back to Roy. Nothing, he told Roy. No alerts. Both men's faces were blank. Roy thanked Ben formally; Ben shrugged, snapped on June Bug's leash, and led her back to his SUV. Her whip-thin tail was tucked between her legs.

Roy kept his own face a blank until the pair had disappeared down the trail. Then he smiled broadly. "I like the way that dog works." He also meant he liked the way the handler worked this homicide scenario. Ben wasn't suckered into finding human remains that weren't there. Neither was his dog.

Roy, a Sevierville optician, and his wife, Suzie—who looks every bit as much the ideal scout leader as Roy does—were training handlers in Eatonton, Georgia, the hometown of *Uncle Remus*'s creator, Joel

Chandler Harris. So it was entirely appropriate that the two of them, with glee and seriousness of purpose, had mustered the equivalent of a tar baby with those mounds of dirt. Handlers, anticipating that cadaver material must be planted *somewhere,* got stuck in their own narrative imagination, talking first themselves and then their dogs into thinking the mounds were graves.

Too many trainings, my own included, proceed in the same fashion: I get to a training site, be it a mildewed house in foreclosure, a harvested cotton field, or an empty, dusty warehouse. Another handler or trainer plants the aids—teeth, a bone, bloody bandages—because it seems pointless to arrive somewhere and not put out training aids. In the missing-toddler scenario, that seeming pointlessness was pushed further. A number of handlers had driven or flown hundreds of miles to attend a National Search Dog Alliance seminar that was almost but not quite in the middle of nowhere. It seemed irrational to come all that way and search for nothing.

The toddler problem was only the first in a long day of Suzie and Roy toying with handlers' minds and challenging their dogs. To start the morning with an elaborate negative scenario may have seemed cruel, but training should make an occasional effort to simulate reality. More than nine times out of ten, Andy Rebmann estimates, nothing is found on actual searches. Police are following vague leads, unreliable jailhouse-snitch testimony, or simply the need to rule out areas. Clearing areas—being able to say, "We don't think the missing person is here"—matters.

The wonderful thing about scenarios is that one doesn't need to invent them. Life itself provides the best material, although re-creating life takes an enormous amount of work, resourcefulness, and, oddly, imagination. Roy and Suzie's mentor, Art Wolff, a Roane County sheriff detective and K9 trainer, developed many of these scenarios for their search-and-rescue team. Many others have benefited from the care and thought that went into these training exercises.

The toddler scenario that Ben and June Bug did so well on was adapted from a 2007 Tennessee Valley case. Police had found a fresh grave and called Tennessee Special Response Team-A. Roy went to the scene with his German shepherd, Cherokee. Cherokee ignored the grave, but police excavated anyway. I can understand their need to know. They found a dead pit bull. That might not have been the end of the story if it had been a larger grave or if Cherokee had alerted. Murderers have sometimes used a dead animal on top of a human victim to mislead investigators.

The scene created that day in Georgia was "just" a training, but that's why setting up realistic scenarios is crucial. Often handlers, like lovers—as fair Rosalind points out in *As You Like It*—can "desire too much of a good thing."

It's not only handlers who let their wishing and wanting lead them into trouble. Their dogs, especially if they aren't conditioned to defeat, can want that good thing just as much. False alerts can have devastating consequences in real life—fruitless days of excavation for law enforcement or charges that prove false.

A grave scenario is especially suggestive. Arpad Vass, whose main laboratories were just an hour down the road from where Roy and Suzie live, noted that our minds have a genius for finding and interpreting anomalies in the landscape. Dogs react in turn, because they, too, excel at finding anomalies and at sensing our reaction. It's not just handlers and their dogs who do this. Geophysicists, botanists, geologists—all of them can look and see what might be a hump, or vegetation that's different, or a suggestive change in the soil—and build an entire grim but fictitious narrative around them.

The negative that June Bug trained on wasn't an entirely blank area, as I realized after watching a few dogs work. Roy and Suzie had scattered odiferous horse-hoof clippings to distract the dogs and irritate the handlers. That was what the dog had munched on while exploring the site. While June Bug had correctly indicated by not indicating, she was a slightly bad dog for grazing on the clippings. Still, she was better than most. Other dogs snagged the horse hooves, but their handlers got increasingly stuck in the narrative tar, committing themselves more and more to a false story as their dogs hesitated and sniffed around the tempting piles of dirt. Several of the dogs alerted on one of the two fake graves. The one that had a generous shot of dead-mouse juice poured across the top.

Mouse juice was Roy and Suzie's serendipitous discovery on a cross-country drive with some bagged and frozen mice in an ice chest that they were planning to use as a training distraction. It got hot. The ice melted, the bag broke, the mice swam. "Holy cats!" Suzie said, her eyes wide with the memory of opening the chest. That's the strongest language I heard her use. The smell must have been staggering.

Roy said, "Cool!" The liquid was even better than the original. He poured it on a mound of dirt. Dozens of blowflies arrived immediately. Mouse juice became another essential element for the elaborate set. Later, Roy and Suzie shared their recipe. I didn't need to write it down. I would remember. Put dead mice in a Mason jar. Add water. Wait a couple of weeks. Liberally pour the libation wherever you want a dog distracted.

It's a sting operation. Done right, as it was in Georgia, it should be reasonably humiliating. As Roy explained to chagrined handlers in their debriefing, slightly buried animal carcass is even better. "Roadkill is phenomenal!" Roy declared. He was beaming. This is a man who, with his wife and partner, experiments on how many hours it takes before dogs alert on fresh blood, or whether incinerated placenta attracts cadaver dogs as much as wet placenta does.

During the debriefing, a couple of handlers tried to make excuses. Roy and Suzie nodded sympathetically, listened carefully, then gently dissuaded them. Even if the dogs were getting chewy treats at bedtime, they shouldn't snack on the job. That's one reason many working-dog handlers wean their dogs off food rewards, although some breeds and some dogs will still work harder for food than toys. Toys can be just as distracting. I've been at patrol dog trainings that included rooms filled with tennis balls in one corner and illegal drugs in another, driving the toy-crazy dogs nuts.

Food, though, unlike purposefully placed dog toys, tends to be almost everywhere on searches, especially at disaster scenes. If a dog is too drawn to food, that can divert precious resources and time. Art Wolff was searching overseas after an earthquake, and one of the dogs on the scene alerted on a spot in a collapsed building. Rescue teams brought in the heavy equipment and started moving the collapsed material. After several hours, they uncovered the refrigerator with rotting food the dog was alerting on. The dog was sent home.

The hoof-and-mouse humiliation was only the beginning that day. Before the sun set in Georgia, Roy and Suzie had played with the minds of a number of handlers in a number of different ways—telling them to stay within a crime-taped area when the wind was bringing the scent of the remains from where they were hidden outside the yellow tape. Roy warned the handlers before they sent their dogs that bear trails were in the area. Bears are notorious for carting off pieces of people.

The dogs kept throwing their heads wistfully toward the scent and getting called back. "She's pulled him off three times," Roy muttered sotto voce, watching one handler urge her dog to come back and search within the confines of the yellow tape, as Roy, playing the role of local law enforcement, had directed them to. One handsome golden retriever, called back a couple of times, finally ran away, ignoring his handler's calls, wagging his tail so furiously underneath a mountain laurel tumbling over the creek bank that the handler asked Roy— standing by sternly, clipboard in hand—permission to search that area. Roy nodded and tried not to show his relief. Please. The dog had found the garbage bag of rich cadaver material stowed under the undercut. What a good bad dog.

That disobedient golden acted just like Suzie's dog, Schatzie, and a teammate's dog did when they were called to try to recover remains from a Russian mob double homicide in Gatlinburg, Tennessee. Handlers were asked to search one area, but the dogs were interested in going elsewhere—where the scent was. Their dogs' unwillingness to stay within the confines of the search area helped the police recover several human parts from an undercut creek bed.

"When your canine shows interest, you really have to trust your dog, you really do," Roy told the handlers in Georgia after they had all worked the scenario. "You've got to follow your dog. Your dog is out there trying to do a job. You say, 'Excuse me, is there any reason I can't search this?' " Then he switched roles from bad cop to good cop because Roy is a sweet man and can play bad cop only so long. "That's a difficult case to work," he admitted.

The scenarios that Roy and Suzie presented that day in Georgia highlight how hard it is to fight human nature. If there are thick briars, officials who have set unrealistic boundaries, or steep rocky slopes, why struggle too strenuously? If your preconceptions tell you that material must be in one area, why go to another? It's one reason you want to bring good dogs in—they're willing to go into hard-to-get-at places. Crime tape and poison ivy mean nothing to them. They're following scent.

The yellow-tape scenario reminded me of the old joke about the cop finding a drunk man crawling on his hands and knees under a street lamp, looking for his wallet. The cop asks him if he's sure he dropped it there, and the drunk says it's more likely he dropped it across the street. "But the light's better here."

The drunk didn't have a dog with him.

. . .

Dogs can't do it all, though. There comes a point where people need not just to "trust their dogs"—as useful and true as the phrase is—but to use and trust their own human brains. Scientists don't know exactly what makes humans' frontal lobes more functional for certain tasks than dogs' frontal lobes, but being able to read Shakespeare isn't the only difference.

If all the search world's a stage, and all the handlers merely players acting out their parts, I think some of us find ourselves in the "Zombie Handler Act." It comes about midplay, often after a riveting dog performance. I've seen it in other handlers and in myself. This isn't about handling zombies but about the danger of becoming one.

New handlers start out as infants at first: overcontrolling, nervous, chattering, mewling, infinitely distracting, not letting dogs do their work and be independent. Good trainers gently guide novice handlers through that irritating phase. "Zip it," Nancy Hook told me.

"Stop hovering over your goddamned dog," Andy Rebmann told another handler.

I got through that phase. Now I was entering an equally problematic phase of dog handling. Because of a couple of recoveries that partly involved being in the right place at the right time, I started to idolize Solo. I knew he could do the work. He was more than five years old—smart, cheerful, fearless, independent, even able to ignore other dogs when he worked, though it was clear he was pretending. Everyone's an actor. Mostly, Solo was becoming likable. He smiled constantly, his mouth open and relaxed, big teeth gleaming. He went into high drive during training and searches, but he cuddled with us at home. After one long difficult search, I said something to David that I will always regret. I was tired, and I shouldn't have said it, even though at that moment, I meant it. "He's my hero."

Solo's success had made me doe-eyed and stupid. He had devoured part of my brain.

When I looked around, I realized I wasn't alone. Zombie handlers were everywhere, made mindless by the fantasy that their dogs were infallible—and could solve complex puzzles by themselves. Experienced working dogs may be capable of certain kinds of problem solving, but it's not their job to strategize. Nonetheless, at seminars and in my own training, I saw people who weren't watching their dogs.

Instead of hovering over their dogs, they now had the opposite problem. They stood on the sidelines, sometimes chatting with other people while their dogs worked.

Dogs do some things much better than humans; other things, not so well. They're much better at scent work than we are, but we don't hand them the car keys and ask them to report back to us at the command control center when they've found what they're looking for. Humans need to set dogs up for success. Dogs need to be put in the right spot to do the job right. That means more than just being downwind. It means partnering with them. I needed to learn when to step aside and when to be helpful to Solo. We were a team. Trusting your dog and letting him do his work doesn't mean being an unthinking chump. You have to keep your eyes and mind open. Nancy Hook snapped at me one day when I was wandering aimlessly in a large field, "You call that a pattern?" Yes. A zombie pattern. I was waiting for Solo to figure out what I wanted.

That's why there is no substitute for watching other people and dogs train. It's only then you fully realize that the stupid things you see them doing, you're doing, too. I didn't have Solo with me when I was watching Roy and Suzie, so I got to observe. I don't know how well we might have done. I've suffered from debilitating stage fright at training, though that has gotten better over time. Paul Martin, running a seminar in Western Carolina, once told me in his slow, comforting drawl, "Your dog is doing just fine, but you're making me nervous."

Some months after my visit to Georgia, I got another chance to realize how much I had to learn. I was in the Mississippi Delta. It was early fall and the cypress, their toes dug deep in the water, were turning gold and crimson; monarchs were wending their way south before the first frosts. And I was getting to learn from Lisa Higgins of Pearl River, Louisiana, one of the many handlers who has trained with Andy Rebmann and gone on to become a top cadaver-dog handler and trainer herself.

If L. Frank Baum had spent his life in Louisiana rather than New

York and the Midwest, Glinda, the Good Witch of the South, might have looked like Lisa Higgins, with large hazel eyes slanting at the corners, a strong nose, round cheeks with slight freckles, and short salt-and-pepper springing hair. Her voice is soft, precise—and firm when needed. When she laughs, which is often, it's a merry peal. Lisa has responded to more than four hundred searches across the United States and Canada since she started training her first dog, Frosty, a golden retriever, in 1990. On Frosty's first callout, in 1991, she helped pinpoint the victim's location, under four feet of water and three feet of sand. Lisa then went on to handle Molli, a Labrador. Now she has Dixee, a wild-child Malinois-German shepherd cross, and Maggie, an aging Australian shepherd who looks like a well-loved stuffed panda. A panda who has helped secure five federal convictions. Lisa has worked with the FBI on numerous cases.

Lisa had set up "a little problem" in Mississippi for handlers who arrived a day early and might want to get started—a simple scenario with some buried placenta. The most basic and wonderful training material. The handlers weren't just to release their dogs to look for scent. Instead, Lisa asked them to focus their search first using something called the "Winthrop Point."

No one in the group had heard of the Winthrop Point. I doubt anyone forgot once Lisa had described it. The point was named after the investigator who realized he saw a pattern in some clandestine grave sites, Lisa told me. Killers, the investigator realized, were doing the same thing that soldiers used to do when they needed to bury excess ordnance. The military needed to know how to get back to it. Murderers want to return to visit their victims but need to recognize where they have hidden their bodies; they also need to know if law enforcement is getting close. The Winthrop Point is a distinctive landmark that won't burn, die, rot, or be covered with kudzu. Trees can't be a Winthrop Point. Gravel roads are out, since roads come and go. A huge boulder might work. A concrete sewer drain. Some permanent fixture in the landscape.

Lisa asked each handler to look around and try to find a nearby point in the growing darkness that might serve as a Winthrop Point. She was standing about ten yards from a large metal contraption with chains buried deep in concrete: a Frisbee golf basket. Distinct. Unmovable. Reasonably long-lasting. Yet only one or two handlers saw what was right in front of them, probably because it was such an innocuous object. Lisa pointed it out and then set them up further: Downhill from the golf basket, the trees were too open for a killer to feel comfortable getting rid of a body. Uphill, though, was wooded, hidden, private. That was where the handler should start her dog. How far back? Lisa reminded handlers that the vast majority of body disposals are less than a hundred feet from a road. These are all facts that handlers need to know. Once the handlers were properly oriented off the Winthrop Point, they could, in turn, orient their dogs.

The next day, it was Lisa's eleven-year-old granddaughter's turn to work her dog. Haylee had just started to train Jayda, an evolving year-old female sable shepherd with maniacal energy. Haylee has an angelic, somber face and soft brown curls. She says "yes, ma'am" and "no, ma'am," especially to her "MaMa," Lisa, who helps homeschool her. It was time for some away-from-home schooling.

"Haylee, you've been listening," Lisa said. "What's the Winthrop Point?"

Haylee had been listening. She rattled off the answer: "Where somebody puts a body where he can find it again. He uses a landmark."

"Why does he do it, Haylee?"

"So he knows when law enforcement is getting close."

"And what else?"

Haylee didn't have an immediate answer, so Lisa gently reminded her. "So he can visit the body whenever he wants. Why?"

Lisa knew that one, too, might be beyond Haylee, so she answered her own question with emphatic precision: "Because he's a sick little puppy, that's why."

Haylee nodded soberly, considered and unafraid. Yes, ma'am. That was plenty enough detail. The human psychology lesson over for the evening, Haylee ran her dog.

When I saw Haylee again, more than a month later, she and her MaMa were in another state. Lisa was training more handlers. Brad Dennis, the charismatic national search director for KlaasKids Foundation, was there as a trainer as well. Haylee was sitting and getting more of an education, taking notes, this time at an evening seminar Brad was teaching, devoted to searches for abducted children and teenagers. Brad has managed search efforts for more than two hundred missing or abducted children around the country. He managed the search after Polly Klaas was abducted and murdered in 1993. He has headed searches during the Super Bowl to stop sex trafficking of boys and girls, an underground Super Bowl perversion. His group rescued six missing children at the Miami Super Bowl in 2010. Brad is one of those ebullient people who knows about the worst, but manages to find and bring out the best in people.

Besides Haylee and her grandmother, who were listening to Brad, the room that evening was filled with tired volunteer and law enforcement handlers. They had already worked their dogs most of the day. I had worked my dog. Haylee had worked her dog. A few of us took notes but most did not. Haylee was an exception; she was writing madly. Homeschooling never ends.

Brad gave the group a scenario from a case that would challenge us. A case he worked that ended, as many do, tragically. In every abduction case, minutes and hours count. A seventeen-year-old disappears while jogging. What should agencies do? Get tracking dogs on the ground immediately. Know exactly where to deploy them. Know how to deploy people around the immediate area of interest. Know the area. Look for trails. Triage where to search first. Separate well-meaning but inexperienced volunteers to the outer perimeter of the search area so that search veterans can concentrate on the high-probability areas.

Brad noted that such work would have been too late for this particular victim. Her murderer had already killed once, molested and raped before. He undoubtedly would have continued. Finding the victim's body led to her killer admitting where he buried his first victim nearly a year before. He had killed one of the girls within an hour of her abduction, the other within an hour and a half.

Brad then cited the grim statistics from a 2006 Washington State study: In 76 percent of the missing-children homicide cases studied, the child was dead within three hours of the abduction. In 88.5 percent of the cases, the child was dead within twenty-four hours. In the majority of cases, 74 percent, the victim was female. Their average age, eleven.

Haylee's hand shot up. She was polite but unapologetic. "Could you go back to the last slide, please?" she asked Brad. Her head was still hunched over her notes. She hadn't gotten down every single percentage. People shifted in their chairs. Brad cheerfully obliged and clicked to the last PowerPoint. Haylee thanked him.

Brad continued with his lecture, talking about "the freeze moment." He showed a grainy bank security video of a girl's abduction that ended with her murder. The girl turned and stood stock-still as her killer approached. That hesitation was all it took; she was gone. All of us, Brad said, have this "awesome, God-given gift." That moment when the hair stands up on the back of your neck. As soon as you feel that? Use it, he said.

Now all the major organizations dealing with abduction have changed their tune on what potential victims should do. Brad looked at the class to see if we knew.

Haylee's hand shot up again. All of us were focused, no longer tired. MaMa, she said, had told her what to do. "If you let them take you someplace else, they will hurt you twice as bad, so you better bite, scratch, kick, and take all the DNA you can."

Lisa beamed. Brad nodded. The handlers clapped.

Haylee is thoughtful and unafraid, smart and studious, curious about the world and all its contradictions. Seven months later, I saw her at another event. She appeared to have grown several inches. She had good news for me. Her manically energetic German shepherd, working in tandem with more experienced dogs, had helped pinpoint a drowning victim.

Haylee is a cadaver-dog handler in training. Even with her great handler genes and the education she and her dog are receiving, she will make mistakes. Everyone does. It's possible that she'll go through a zombie handler phase. I expect it will be brief. Her MaMa won't allow it to last.

12

The Grief of Others

Paula Bronstein / The Hartford Courant

■ Connecticut state Trooper Andrew Rebmann and his dog Josie walk in the woods along Route 140 near New Bedford, Mass., Tuesday in a search for the bodies of two victims believed to have been slain by a serial killer.

■ Rebmann, center, goes over a map of the search area. With him are District Attorney Ronald A. Pina, left, and state Trooper Kevin Rodino.

Police dogs seek victims of killer

By LYNNE TUOHY
Courant Staff Writer

NEW BEDFORD, Mass., — Six German shepherds and their handlers scoured wooded areas along three major highways Tuesday in a grim, and possibly final, search for victims of a serial killer.

The dogs ran through swampland that came up to their bellies and became entangled in dense patches of briar that resembled snarled barbed wire. Collectively, they found a discarded wallet, a dead dog in a bag and countless wild animals killed by automobiles — but no victims of a killer.

The three-day search was planned and coordinated by Connecticut's leading trainer of search dogs, state police Trooper Andrew Rebmann. The searchers in this working-class community east of the Rhode Island border are looking for at least two bodies. If found, they would be added to the eight already discovered along routes 140, 195 and 88.

"I think the big thing about this is the scope of the search, the amount of area that has to be covered and the potential number of victims," Rebmann said.

All the known victims were drug addicts, and some were prostitutes, according to law enforcement officials, and they were known to frequent a dingy area of the city known as Weld Square.

"It's almost like a McDonald's window," said Bristol County District Attorney Ronald A. Pina. "It's a fast-drug thing."

Pina will not say how the victims were slain or the positions in which they were found, except to say that there is a pattern in both areas.

"I'm holding that back so that only the killer and we know," Pina said. Several people who have confessed to the slayings

See Police, Page A11

And the bones sang chirping
With the burden of the grasshopper . . .

—T. S. Eliot, "Ash Wednesday," 1930

There are only two options. Except that one of them gives rise to a third option. The one that gives me nightmares.

The first option is as close to ideal as you can get, given that you are looking for dead people: You and the dog help find someone or part of someone. The person's disappearance and death is usually sad, sometimes tragic, and sometimes not so sad. For me, it's not nightmarish to find someone. It doesn't create post-traumatic stress. It represents success. I celebrate, but quietly. I understand when people make mournful faces and suggest that it must be rewarding but certainly not enjoyable work. For me, it's both. What Solo and I do isn't dutiful volunteer service. I train the dog and myself because, first and foremost, it's fun. I don't dread getting called out. I hope we are: It's a challenging puzzle that pushes both of us to our mental and physical and scent limits. Plus, I get outside, often in the woods, and I can watch Solo use his nose—one of the most pleasurable sights on earth. I hope we find the person we're looking for.

Option two is not finding someone. Speculative searches are part of the job. Most searches end with no victim found. Option two haunts me more than option one. My mind keeps poking at the possibilities, wondering about the weaknesses in our search strategy, in the dog's work, in my work, turning over other options, seeing if they resonate, testing a theory, discarding it, picking up another and feeling its contours. Inevitably, option two involves trying to imagine what happened at the end for that victim, and where, and how.

Option two can stretch out for years. Or longer. Nonetheless, it, too, falls within the realm of normal. Not finding someone happens all the time. It's not just in underdeveloped countries that missing people stay that way. In the United States, the list of "endangered" people—law-enforcement-speak for highly likely dead—had nearly forty-eight thousand entries in 2012. That's a medium-sized city filled with no resolution.

The grief and nightmares of not finding someone belong to the missing person's family and friends, not me. It would be presumptuous to appropriate them.

My nightmares about searches—the ones that rightfully belong to me—rise from option three: if I ever find out that we missed someone or something in an area we were responsible for searching. I know a number of handlers feel the same way. It's our special dread. As much as I hate an occasional false alert, I hate a false negative even more: where a dog ignores or accidentally overruns scent that's out there. It can happen. To make option three even worse, unless the remains of the person you were searching for are found—and found at a comfortable distance from the areas you searched—you never entirely know.

Sure, I trust the dog; I also like to verify. I don't always trust me. Or the terrain. Or the search conditions, which are never as easy as the most difficult training you can set up.

Each difficult search provides lessons and inevitable Monday-morning quarterbacking. Did we miss that bone? Should I have insisted on reworking the area that was already cleared with Bush Hogs and a multitude of line searchers and at least one other cadaver dog? That area where the smell of decomposition was so thick that every time a semi drove by, it would kick up a waft of sweetness that covered our clothes and got up in our nostrils. Where every dark oily spot of killed vegetation, "body burn," was accompanied by the fur and bones of animals. What about that little bone lying next to that small animal skull? Was it something other than animal? Should I have brought Solo out of the car then and there and run him over those dozens of skeletons to let him do a preliminary sorting? I didn't. I did what I was told. And it was only in the days after that I started stewing and rethinking each and every moment.

I waited for our next assignment. And the next, going up the road to search a barn and abandoned house. Then we searched around the pond at the top of the hill. And then we went on, following exhausted investigators up a rutted clay lane to a mobile home in a hollow, with an old SUV sitting outside. It had taken us three tries to find it. We kept backing down one-lane roads because we couldn't turn around without falling off the verge and into trees. Finally, there it was. Finding the place felt like a small victory.

The woman who had called the investigators, worried about a bone that one of the dogs had brought home, came out of the trailer and pulled her barking dogs inside and shut the flimsy door behind them. I could hear them snuffling on the other side. The dogs, she said, were always dragging something in. I looked at a recently dead vole just to the left of my boot, its fur matted with dog saliva. That's what dogs do. They bring stuff home. She followed my gaze and shrugged in apology.

You just never know, she said. She paused and went on. I kept thinking and thinking about it. You just never know. I couldn't live with that. One investigator assured her she had done the right thing in calling. It wasn't her job to sort out human from nonhuman. Her preadolescent daughter was standing silently behind yet alongside her. She looked like her mother but without the tiny silver rings on every finger and along the earlobes. Not yet fully pierced. Silent, fey. I knew what her mother meant. Her mind had hooked on a detail, caught in the threads of "what if?"

The bone was sitting on a wood post. It looked to me like a vertebra from a mammal no bigger than a possum. It had a fleshy pinkness just underneath the gray whiteness. It had decomposed for a couple of days. I looked at the investigators, then used a long stick to snag the bone through the hole where a small spinal column once ran, and took it over to a dusty, clear area of the yard and liberated it from the stick. I wasn't disturbing a crime scene; the dog had already done that by toting the bone home from wherever he'd found it. I took Solo out of the car and ran him past it. He caught the scent, hooked around when he smelled the decomposition, sniffed briefly, then moved on. Not the kind of decomposition that would garner him a reward. The investigators and I took Solo to the back area of the property, where the dog had emerged with the bone. He ran the area quickly, without changes in his body language, without looking at me, working the edges back toward the car. The woman thanked us. We thanked her. For caring.

Next, we searched around a pond. Nothing. A trail to a deer blind. Nothing. A barn. A spot where hunters dump deer parts. A mattress covered with stains on the side of the road. Side roads with piles of trash at the end. Every white garbage bag, every black garbage bag. Hand sweep. Check here. Check here. Check, check. Time search started: 10:03 A.M. Time search ended 10:17 A.M. Time started: 10:42 A.M. Time ended: 11:22 A.M. I gave Solo a break with water and a full blast of air-conditioning until his tongue stopped dangling sideways off the shelf of his jaw.

Down the road to the next pull-off to start the clock all over again. No interest. No alerts. Break for cold chicken sandwiches and soda. For coagulated greasy pizza and bottled water. Back to work off gravel roads. Checking drain pipes. Going down into the creek running parallel to the road. Over the deadfall along the tree line.

This is the reality of searching: You cannot see the world in a grain of sand. It's the opposite. The grain you are looking for is so infinitely small, so lost in the world, that it might never be found.

And yet it wasn't all mournful. Toward the end of one day, we watched Solo, who should have been exhausted, levitating through the high grass, clearing a final area before quitting time, bouncing like a black-and-red India rubber ball, backlit by an early-evening sunset. He made us smile. A happy shadow who goes out in front of me. I drove home so tired that even the tickle of a tick on my neck elicited only a flick from my finger. Solo, dried mud flaking off his guard hairs, was sacked out on the backseat, not a whine left in him.

Later, when I learned that searchers had found human remains in an area that Solo and I hadn't searched, I didn't care that we weren't the ones who had located them. All I felt was a gut-wrenching relief that remains had been found at all. It was a deeper, selfish, and utterly prosaic satisfaction to learn that I didn't have to continue worrying about option three for this case any longer. I could cross it off my list of nightmares. We hadn't skipped over anything. Solo's big nose hadn't gotten close enough to stand a chance. After I got off the phone with the kind investigator, I pulled to the side of the road. I just sat for a while, until I could breathe steadily again.

. . .

North Carolina doesn't feel dangerous to me. It shouldn't. It's much safer here than it was in the 1970s. The murder rate, like all serious crime, has dropped precipitously since the 1970s and 1980s, when poverty and crack ruled—more than 60 percent. Poverty and addiction still take their toll, but their cut isn't as deep. Statistically, I'm very safe. I don't have an abusive partner or parent. I live in a decent neighborhood. We don't have guns in the house. I don't need to sell my body to feed a drug habit. We have a noisy Irish setter in our house who devilishly encourages the German shepherd to be noisy as well.

Nonetheless, my relationship with my surroundings has changed since I started working with Solo. I no longer watch turkey vultures gliding in lazy circles, especially if more than three stack up in the same thermals, without wondering whether they're smelling something more than a white-tailed deer carcass far below. We keep track of vultures on searches, although it doesn't take much to attract them. One day I watched four on my urban street competing over one squashed squirrel, clumsily landing on a neighbor's tarpaper rooftop before swooping down to squabble over a couple ounces of protein.

It's not just in the woods that my viewing habits have changed. I used to avoid and scorn the top of television news and web news, with their insistent focus on violence and crime—the cheapest, easiest thing to cover and get high ratings. Now I tune in quite purposefully if someone is missing. Then I obsess about whether I'll get a callout on that case. Why bother training otherwise? But with a couple of rare exceptions, when they ask for volunteers, I don't call the police. I wait for them to call me. That doesn't keep me from wishing and hoping.

When the burden of not acting becomes unbearable, I'll call Nancy Hook, my equivalent of a twelve-step sponsor for this compulsion, so she can remind me what I might lose by calling law enforcement: my dignity and self-respect. "You're not an ambulance chaser," she'll tell me sternly. "You're a professional." We'll chat, she'll make me laugh, she'll tell me she has to go feed the horses, and I'll remember that I've got a curriculum committee meeting I'm almost late for.

After the meeting, if the itch returns, I can always channel Andy Rebmann's stentorian voice: "You do not self-deploy," he said, slowly emphasizing each syllable, glaring at search-and-rescue volunteers at a seminar. I've seen SAR-team self-deployment. It's not pretty. Nancy and Andy are right: It has the same scuzzy feel as personal injury lawyers—the kind who advertise with 800 numbers on late-night television—showing up at the scene of a wreck.

In my defense, the thinking I do about missing persons cases isn't entirely wasted. Even the callouts that never come can add to my knowledge base: Google Mapping how to get to the area if I'm called, thinking about winds and temperatures and humidity over past days. If the area has been defined in the news, I stare at the satellite view, look at the dents in the vegetation, wonder if they represent a creek or a trail.

One also has to be prepared for what a body might look like. Andy has a slide show that provides a whirlwind tour: There's rigor mortis, putrefaction, skin slippage, and liquefaction. Jay damage. Crow damage. People tied up, burned up, pulled up from lakes, crushed in disasters. Scattered by bear and coyote. Handlers need to have a realistic notion of what bodies look like after a few days, weeks, or months out in the environment. A search is not an academic exercise. Bodies are never pretty in early stages; later, they can fade into their environment like camouflage. It's critical to be able to recognize a spot where the soil has turned so acidic from a body lying there that plants die.

Soon enough, the bodies deflate and fade into the North Carolina woody foliage, a slightly darker or sometimes yellowish leathery accent under the dark green. You would have to know where they are to see them. Or have a dog around who can tell you.

On a recent search, the detective flanking me asked me whether cadaver dogs can miss or skip over a body. It depends. I looked around at the impenetrable woods on one side and the clear-cut mess on the other—logs lying crisscrossed, shrubby growth coming up in between. A swamp lay behind us; suspicious tire tracks were visible along the dirt road. We had punched in several places where slight deer paths, or even a break in the vines and undergrowth, provided a gap. The working presumption, not a stupid one, was that someone trying to carry a body wouldn't have an easy time, either. If Solo hit scent, he would follow it if he weren't too exhausted and panting to bring in scent. But his nose had to be in the proper place. Getting his body levitated over impenetrable brush isn't that easy. Hasty searches over dozens of acres don't give 100 percent coverage. People talk about grid searches or line searches in a casual way, but in many areas in North Carolina, doing that would take a phalanx of Bush Hogs running in front of you. This was triage. Everyone does his or her level best—a bit of whacking with a machete when it's feasible, saving a bit of energy when you can see a spot for entry just ahead.

You try to maximize the odds by knowing as much as you can. Before I tackle a new kind of search—say, an Alzheimer's victim who wandered off, someone who was separated from a violent husband, or a drug user who was desperate for a fix—I will go back and hit the research.

Take a person with Alzheimer's or dementia. Her behavior differs from that of other lost people. If an unimpaired person is right-handed and gets lost, she tends to move to the right. Not someone with dementia. That person doesn't behave logically, even at a subconscious level. She will walk straight into thick brush. Her brain can't compute turning around and backing up. She will keep walking in place. The body of a man with Alzheimer's was found in the woods two streets away from his suburban home, a month after he went missing. The police, I heard through the grapevine, had been given my name and number soon after he went missing. But the call never came. There could be a hundred good reasons they never reached out. Or none. It was one of those times when I sat and waited.

Most people with dementia or Alzheimer's—nearly 90 percent—are found less than a mile from their point of departure and within thirty yards of a road. I know that because, when I was waiting for the call, I did the research. It wasn't wasted effort. I've used that knowledge for other searches. I still think about that one man and his family. Perhaps even more than if I had gone on the search.

. . .

How do you search properly along a roadside? How many yards back would your basic panicked or lazy murderer drag a victim? Not far. Twenty-five yards. Check farther back. Animals are more industrious: How much farther back might they drag something? Much farther, depending on the animal. Dogs have been known to carry parts of people up to a couple of miles. What kinds of animals inhabit the search area? Bears tend to go downhill to a clearing or creek; they can move a whole body, crack femurs, and bat heads around like soccer balls. Coyotes can go uphill to a den, carrying limbs. Possums and raccoons tend to dine on the spot, although they can drag material vertically. We have coyotes in every county in North Carolina. We have black bears in 60 percent of North Carolina, throughout the mountains of western Carolina, and down through the swamps and shrubby pocosins of what we call "Down East."

Then there are the obvious areas that need searching, as Brad Dennis pointed out: abandoned properties and outbuildings. Piles of wood and debris that can be used to conceal a body. Impromptu garbage dumps where someone can drag an old mattress over the body. For clandestine burials, natural holes made by roots and erosion that form ready-made graves, with only minimal additional digging needed. Did the suspect have easy access to a shovel? Wasn't he homeless? Most clandestine burials are no deeper than two and a half feet, yet that's enough for someone to disappear forever. Arpad Vass calls the clandestine burial his "nemesis."

There's the time frame to consider. In North Carolina, areas can get overgrown in one season. Hunters tend to find skeletal remains more often than law enforcement officers do. Mostly skulls, as they are the easiest to identify. A turkey hunter found a skull more than a year and a half after a young girl disappeared. The search for her remains was one of the most thorough mounted in recent memory in North Carolina. Other bones tend to blend in with leaf litter like chameleons. But anyone who has searched in North Carolina woods knows that heart-stopping moment when you see a light brown or green-moss-covered turtle shell, a hump coming out of the surrounding humus or leaf fall, and momentarily mistake it for a skull.

On one case, police jokingly—but with an underlying awareness of the neighborhood surrounding the woods—asked me to please find only the body they were looking for. On another case, searchers found skeletal remains, but not the victim they were looking for. One of the medical examiner's investigators explained that this was simply part of the business. She and other forensic investigators, she said, can't see a black garbage bag in a ditch along a road without wondering.

Nonetheless, for all the cruel casualness of people and of nature, there's something reassuring about working a cadaver dog. It's true that finding someone or part of someone can give closure to a family or allow the police or prosecutors to move ahead with a case. That doesn't entirely explain why it's important to find remains, even if there's little to nothing left. It's partly to be able to acknowledge, even momentarily, the spot where someone was hidden or dumped. And to think on it. I like the fact that, animal predation aside, it can be hard to get rid of a body. I love the fact that when people die, they don't completely disappear, despite their murderers' efforts. Yes, they cease to exist. At the same time, they also stubbornly stick around.

During one search, Solo went right to a spot in the woods, lay down, and looked at me expectantly. An investigator confirmed Solo had alerted on the exact spot where more than a year earlier, hunters had found the bones of a murder victim. The pine forest floor held on to her scent and would do so for years.

· · ·

Such dreary streets! blocks of blackness, not houses, on either hand, and here and there a candle, like a candle moving about in a tomb.

—Herman Melville, *Moby-Dick*, 1851

Three main highways go in and out of the old whaling town of New Bedford, Massachusetts: Interstate 195, state Route 140, and U.S. Highway 6. During the late 1980s, the height of the crack cocaine and heroin epidemic, those were the highways used to ferry drugs in. Those were the highways where women's bodies were dumped on the way out of town. During that same time, one reporter noted, a local clinic was treating four hundred heroin addicts a day. Only Boston beat that number in the state. Now, in New Bedford, as across the nation, crime is down, way down.

But for six months during that epidemic, from April to September 1988, eleven women, most of them desperately selling sex in exchange for cocaine or heroin, disappeared from Weld Square, a dreary block of darkness in the center of town.

One woman's body was found while other women continued to disappear. No one connected the cases until it was too late. These were women whose lives had started to slip away before they were murdered.

In early July 1988, a woman stopped her car along state Route 140 to pee in the nearby scrubby brush. She discovered the first body. Debra Medeiros, twenty-eight, was spread-eagled, her bra wrapped around her neck. She had been missing since late May. Later that month, two motorcyclists also felt the call of nature and found Nancy

Paiva, thirty-one. Paiva was on her back, her feet pointed toward the westbound traffic on Interstate 195. Next, a public works employee collecting cans on his lunch break found a third woman's remains: Debbie DeMello's body was just off an I-195 onramp.

That was when the Bristol County district attorney's office contacted Andy Rebmann's supervisor at the Connecticut State Police. By that time, Andy was working Lady's replacement, Josie—another Fidelco dog who wasn't cut out for guide-dog work, just like Rufus and Lady. Too much drive. She was cute, light on her feet, intense. Not that big for a shepherd and as tightly articulated as a cat. Andy hadn't been working her long, but that didn't seem to matter. She was a natural, cross-trained to find both live people and dead ones. She didn't care which as long as she got her reward. She was the kind of dog who would dash two or three times into and out of impenetrable brush to find Andy—to make sure he understood, hitting his pocket with her nose. The ball. The ball. The ball. Jeez. Get it out already. Her first callout, the day after she was certified, was for a suicidal person. It took her two minutes to find the guy. Still alive.

"She was a lot of fun," Andy said simply. "She was the easiest dog I had ever trained in my life."

The New Bedford highway search was not fun. It was dangerous and hard going. Dense traffic on one side. Claustrophobic thorns, brush, pine, and dead animals on the other. Josie worked for five hours that first day, searching the north side of Interstate 195. That may not sound like a lot of time to people who punch in and out and get to play on the web for part of the workday. But for a search dog spending all her time sniffing and quartering and leaping over obstacles and getting caught up in dense brush, it's a brutal schedule.

Josie was young, though, three years old. And while Andy wasn't a spring chicken, he was fit and experienced. Andy set up half-mile sectors and worked the shoulder. Then he'd go in twenty-five yards and work inside the deer fence. Nothing. All that day.

Nothing the next morning, either. By midafternoon, Josie and Andy had worked their way down to the ramp coming off Reed Road. The north side. They would have to do the south side, but all in good time.

Like most operational air-scent dogs, Josie was off-lead so she could go where her nose led her. Suddenly, she was in the trees, not twenty-five feet off the ramp, tail wagging madly. She bounded out of the woods and hit Andy's pocket. Give me the ball.

It was the remains of Dawn Mendes, twenty-five, from New Bedford, last seen leaving her home on September 4, 1988. Josie found her on November 29, 1988. After Mendes was identified, the New Bedford *Standard-Times*'s headline was blunt and offensive. The headline started with Mendes's body, went to her sex work, and left her name out entirely: I-195 BODY IS CITY PROSTITUTE'S.

Andy gave Josie a day's break and then went back to searching on December 1, 1988. Back and forth. Debbie McConnell, from Newport, Rhode Island, disappeared sometime in June 1988. Josie and Andy found her in the midafternoon on December 1, 1988, down an embankment off Route 140 northbound, thirty feet from the road. McConnell was less than three miles from where the first victim was found.

These were linear miles of demanding work, going twenty-five yards in, coming out, gridding the length, working the high spots, trying to keep the dog downwind, mostly avoiding dead deer and smaller mammals. Josie did manage one delightful break where she rolled in a dead skunk. At one point, a television truck crew distracted the sociable dog, and she started to dash across the highway to greet them. A semi barely missed her.

Andy and Josie were back at it in late March 1989 when Josie found a third victim: Robin Rhodes, twenty-eight, off state Route 140 southbound, lodged within the trees, just twenty-five feet from the highway. After that, Andy organized a four-day search, pulling in six dogs and their handlers from four states. They found no more bodies. After several days without results, Andy called a halt to the search. "At least we know where the victims aren't," he told a newspaper reporter.

Only one more victim was found after that—Sandra Botelho— miles away from the others, on I-195 in Marion. Nine women's bodies in nine months. Two other women who fit the profiles of the Weld Square victims remain missing. Although the police honed in on two suspects, a third possibility emerged: The killer might have worked on a seasonal fishing vessel based in New Bedford.

Lighthearted Josie died in 1991, two years after that search, of a blood disorder. She was still a young dog. Andy retired from the Connecticut State Police that same year.

If Andy hadn't pushed forward, would as many women have been found? I didn't bother asking him; I already knew the answer. Unlikely. This was not a case where hundreds of good-hearted volunteer searchers showed up and the Red Cross arrived to provide Gatorade and Dunkin' Donuts for everyone. The case didn't become a case until it was too late for five of the victims. Would it have changed anything for the women if they had been warned that a killer (or killers) was targeting them? Perhaps not. No one will ever know.

In New Bedford, a small number of people—and one dog who loved to work and play—cared a lot. Yet all of the victims, except the three Josie and Andy found, were found by accident: people needing to relieve themselves on the side of the highway; hunters; two boys walking along the breakdown lane and seeing a body just ten feet away.

Twenty-four years later, the cases remain open and unsolved.

13

All the Soldiers Gone

You smug-faced crowds with kindling eye
Who cheer when soldier lads march by,
Sneak home and pray you'll never know
The hell where youth and laughter go.

—Siegfried Sassoon, "Suicide in the Trenches," 1918

Dogs have been used during war for centuries: first for attack, then for scenting the presence of enemy soldiers, then for sending messages to the front, and then for finding bombs and mines. Starting in the nineteenth century, dogs were also tasked with finding wounded soldiers. Through it all, the dogs' stated purpose in war was to focus on finding the *living* rather than the dead. Dogs were supposed to assist in the heat of battle, not find the bodies in its aftermath. Canines were simply additional cogs in the machinery of war.

Nonetheless, immediately after a battle, triaging among the dead and the living is a critical task. The Red Cross knew that better than any organization. World War I saw the first widespread deployment of the dogs Germans had started training in the late 1800s: *Sanitäts-hundes,* or "sanitary dogs."

These dogs carried saddlebags filled with medical supplies that wounded soldiers could remove once the dogs arrived at their sides on the battlefield. The Germans used German shepherds; the English used a mixture of breeds from Airedales to collies to mutts. All of them became famous as "mercy dogs" or "ambulance dogs." At one point during World War I, the Red Cross estimated that ten thousand dogs were working on both sides of the front. While their existence and their help are well documented, the skeptic in me looks at that suspiciously round number with a bit of distrust—it makes me wonder if the Red Cross PR machine was working overtime a century ago.

Mercy dogs dealt only with the living. That was a position that organizations like the Red Cross maintained, even throughout World War I. Despite the combat saying "Leave no man behind," on the battlefield, in the fog of war, the dead are left. In war, we create piles of dead. In "great" wars, like World War I, where Siegfried Sassoon was awarded a medal for bravery and bitterly witnessed his comrades obliterated in muddy trenches; in "good" wars, like World War II; in wars that we admit were disasters. During World War I and World War II, hundreds of thousands of soldiers were left where they fell or were thrown into mass graves. Tending to the living and wounded was already an overwhelming task. That hasn't changed—on disaster scenes today, we prioritize finding the living before trying to recover bodies.

"Dogs are never trained to scent out the dead," wrote Ellwood Hendrick in a 1917 issue of the *Red Cross Magazine.* "Their business is to assist the wounded." He went on in a vain effort to dampen the patriotic fervor that the dogs encouraged: "We do not have to go about killing people to make dogs worthwhile."

The issue of recovering the war dead had long been a central concern. In the United States, the first official efforts to recover soldiers' remains were after the Seminole Indian wars of the early 1800s. That program was highly ineffectual: The laws provided no funding from the government to pay the expenses to return a dead family member. As Michael Sledge noted in his marvelous history, *Soldier Dead,* relatives could have the remains shipped back if they provided a lead-lined coffin to the "designated Quartermaster at a port."

Few families could afford the coffins. Nonetheless, trying to recover soldiers—at least those on the winning side whose families were wealthy—had begun in earnest. By the Civil War, Sledge notes, the country started "honoring the death of the common soldier." Although the war dead in that conflict became a logistical nightmare, the federal government stepped in. By 1873, more than three hundred thousand soldiers had been reinterred in seventy-four new national cemeteries. Though the total number of dead on each side of the conflict is still a mystery, the national cemeteries were reserved for Union soldiers. "The absence of official concern for the Confederate dead stood in stark contrast," wrote Drew Gilpin Faust in *This Republic of Suffering*, her recent history of death in the Civil War. An *Atlantic* magazine writer on a tour of Virginia shortly after the war, Faust noted, came upon two bodies in the wilderness. His guide examined their uniform buttons. "They was No'th Carolinians; that's why they didn't bury 'em," the guide informed a horrified John Trowbridge.

Civil War historian J. David Hacker now thinks that earlier estimates of 620,000 men dying on both sides—a figure that has stood since the nineteenth century—was far below the actual number. His most recent work puts that number at 20 percent higher: that the war was responsible for more like 750,000 men dying.

But what do these numbers mean? Why do they matter? Partly, as

Hacker, an assiduous and brilliant historian, notes, "it's our duty to get it right." His new estimate suggests that more men died as a result of the Civil War than from all other American wars combined.

And yet all too often we count everything—and understand nothing.

In the Pacific Theater during World War II, journalist Ernie Pyle drafted what was to be his final column. It was found in his pocket when his body was recovered on the island of Ie Shima on April 18, 1945. He had been shot by a Japanese machine gunner: "Dead men by mass production—in one country after another—month after month and year after year. Dead men in winter and dead men in summer. Dead men in such familiar promiscuity that they become monotonous. Dead men in such monstrous infinity that you come almost to hate them. These are the things that you at home need not even try to understand."

. . .

E pluribus unum. Out of many, one.

Although there were scattered reports of Israel using dogs during the Sinai War in 1973, it took until 2003 for the U.S. military, which had been involved in the earliest experiments on the feasibility of cadaver dogs, to use them to recover dead soldiers. Even then it was a modest yet controversial experiment—to send just one man and his two German shepherds from the Northeast into the jungles of Vietnam to look for MIAs who had disappeared decades before.

The military sent Rhode Island state trooper Matt Zarrella. It was February, and Vietnam was hotter than Rhode Island. Matt's nine-year-old female shepherd, Panzer, was panting. His one-year-old male, Maximus, was panting. As well as they could. The nervous Vietnamese pilots had insisted that the two dogs be muzzled before they climbed into the helicopter to fly toward the southernmost tip of Vietnam.

Now they were over rice paddies where the craters from carpet bombing from decades before were still visible. The temperatures inside the old Soviet helicopter were rising with the humidity of the jungle air. Matt wasn't muzzled—although with his propensity to speak his mind, that might have been a good idea.

The mission to find MIAs had added urgency. Witnesses to fighter jets that had been shot down during the height of the Vietnam War, in the 1960s, were dying out. The recovery efforts had slowed, with fewer and fewer U.S. soldiers' and pilots' remains found. That was when senators John McCain and Ted Kennedy, impressed by dogs working at the three 9/11 disaster sites—Shanksville, Pennsylvania; the World Trade Center; and the Pentagon—went to the military. Why weren't cadaver dogs being used in Vietnam to find servicemen who had been missing more than thirty years? The idea was far from universally embraced. It went forward nonetheless.

Matt Zarrella and his dogs had a reputation in the Northeast for finding the dead and the buried, so he got the call one day. Would he consider consulting with the military unit that oversees the recovery of servicemen—the Joint POW/MIA Accounting Command—about whether dogs might be effective?

The retired military officer assigned the job picked Matt's brain. What would they need to get such a program started? What would the handlers need? The dogs need? The conversation moved in an obvious direction.

"It just came down to asking me if I would do it. I was extremely honored," Matt said. He would become the first cadaver-dog handler to officially search for missing soldiers from any U.S. conflict. When Matt flew to Hawaii to prepare himself and the dogs at the JPAC Central Identification Laboratory, 1,889 servicemen were still missing in Vietnam.

Matt, a former marine himself, was a believer in the credo "Leave no man behind." Nothing, though, is ever that simple. The repatriation of soldiers' remains from Vietnam had been a disaster during the

1970s: politically fraught, expensive, often based on unreliable information, and dangerous.

Two years before the fall of Saigon, a field team trying to recover deceased U.S. servicemen was ambushed. The team's leader, Captain Richard Reese, tried to save his men by standing up and telling the Vietcong they were unarmed. He was killed in a fusillade of bullets. For the rest of the 1970s and until 1985, the United States made no more efforts to bring the dead home.

Even when the effort resumed, some of the recoveries were spectacular failures. One, at Yen Thuong in central Vietnam, was based on faulty information: It was probably a missile site or a downed Soviet plane, not an American plane. Military anthropologists excavated in any case. They dug a hole that was fifty by one hundred feet, and forty feet deep, sifted through all that removed clay, and found nothing.

Matt, who knew some of the political problems, couldn't help wondering if he was being set up, consciously or unconsciously, to be the fall guy for this first military cadaver-dog project. He knew holes had been dug all over Vietnam to no avail. He had already witnessed the tensions between anthropologists and military personnel about whether he and his dogs were capable of finding human remains, or whether Matt simply had canine versions of Clever Hans. There was an argument on the military base in Hawaii over whether they would have to kennel and quarantine his search dogs for six months because of rabies regulations if they left the base, but the base had no reasonable kennels. The dogs ended up staying in a tiny retrofitted closet. Once Matt got to Vietnam, the vermin- and spider-infested hotel in Ho Chi Minh City didn't improve his mood. He didn't mind tough conditions and carrying a fifty-pound backpack, but he was forty-one. Not as old as his female shepherd in human years but no longer a young man.

Matt also wondered whether he was setting up his own dogs to fail. He knew that Panzer could find the dead and the buried: She'd done it for years in the Northeast United States. But the oldest buried remains she'd found were a decade old, not going back nearly four decades, as they would be in Vietnam. The temperatures and conditions were not only unfamiliar, they were grueling. Panzer could work in snow and ice, but mangrove swamps and rice paddies with 100 percent humidity and heat indexes above one hundred were new territory.

Matt talked to Andy Rebmann. Andy considers Matt the equivalent of an adopted son, and Matt adores and respects Andy. They both have an ability to argue. They argued about whether Panzer was too old to go. Gunner, Matt's middle-aged Swiss mountain search dog, had gotten cancer and had a leg amputated after Matt committed to going to Vietnam. Matt couldn't back out; he'd promised two dogs. So he rushed to the pound and found a six-month-old German shepherd. The adolescent shepherd had been labeled aggressive. He wasn't; he had drive. Matt called him Maximus and trained him hard over the next six months—but the one thing he couldn't do was make Max age faster.

So Matt was left with a one-year-old dog and a nine-year-old dog—positioned at the extreme ends of the age spectrum for search dogs. One unproved, just certified, and the other old and soft enough to tire quickly or even die. A bunch of skeptical anthropologists in Hawaii and Vietnam were keeping track. Not a warm and fuzzy atmosphere.

"I tried to explain. We're not here to replace you. We're just another tool. We need good investigative tactics. We're only a small portion of a team," Matt said. He was secretly worried that his two "tools" might not be operational.

Captain David Phillips's fixed-wing fighter jet had been shot down on July 3, 1966, in a thicket of mangrove trees on the southern tip of Vietnam. A witness told authorities that he had recovered Phillips's remains and buried them. Others had reburied him. The jet, if there had been pieces left, had been entirely repurposed. It was acidic soil. Bones would disappear—if they had been there in the first place.

The search went quickly once they descended. Panzer alerted on the exact spot that the villager said he had buried the remains near his home. Fine. The anthropologists were planning to excavate there anyway. Despite a preference to search blind, Matt couldn't help thinking that might not be the end of the story, "just based on the profile." So he and Panzer wandered toward the back of the house and the thick jungle at the outskirts of a village, toward an old family cemetery and garbage dump about 150 feet away from the site that was supposed to be ground zero.

That was when he noticed Panzer's body language start to change. She eyed Matt, threw her head, worked a small area hard, but didn't give a final alert. She was getting tired. He put her up and got young Max. Max did the same thing Panzer had, in the same area, except Max gave his final alert. "He was sure it was there. It was in his eyes."

Matt put Max up and went to talk to two of the anthropologists, who were intrigued. Matt and the dogs flew to other sites. Some would be burials; some would be crash sites; some would be spots where prisoners of war reportedly died in camps and were buried by villagers; some would be entirely invented. Matt would come in, search, and the dogs wouldn't show any interest. Nothing here. The villagers would look at Matt's dogs and changed their stories about remains being there, probably to what was closer to the truth. The dogs hadn't been trained for this new task, but apparently they were good lie detectors.

Nearly a month later, Matt was in the lobby of his Ho Chi Minh City hotel and ran into one of the anthropologists from the Phillips site. Matt and the dogs had just finished their last case. They'd found nothing tangible. Matt's morale was rock-bottom.

"Did anyone tell you what we recovered at the site you searched?" the anthropologist asked. Matt had heard nothing. The anthropologists had excavated the general area where Max had alerted and Panzer showed interest. Six inches down, they'd found a pocketknife. A zipper from a flight suit. Pieces of life support equipment. And what the anthropologist thought was a human patella, a kneecap, although that hadn't been determined. The term "osseous remains" was the one that stuck.

"I wanted to break down and cry," Matt said.

A couple of years after that, having heard nothing more, Matt decided to pick up the phone and call the Identification Laboratory in Hawaii. He got a sympathetic guy on the other end who looked up the case. Captain Phillips's remains had been identified in September 2004. Panzer had died of cancer in September 2004. But Matt now knew the rest of the story. Maximus went on to many more searches. In 2011, he, too, died of cancer, at the age of ten.

Today, 1,664 Americans remain unaccounted for from the war in Southeast Asia, 225 fewer than when Matt went over with his dogs. Three hundred thousand Vietnamese soldiers are still missing. The military has not sent another dog-and-handler team back to Vietnam.

Captain David Phillips's wife died in 1989. His daughter Debra Stubbs went to the military laboratory in Hawaii to bring back her father's remains, wrapped in a military-green wool blanket, according to news accounts. She slid her hand inside, she told an *Atlanta Journal-Constitution* reporter. That was the closest she ever got to her father, who had gone to Vietnam before she was born. Her mother, she said, had worried for years about whether her husband was a prisoner of war. She kept telling the family she was going to get on a plane to Vietnam to try to discover the truth for herself. She never did.

Captain Phillips's three daughters and his brother buried his repatriated remains in the Bonaventure Cemetery in Savannah, overlooking the Wilmington River—a cemetery filled with live oaks strung in trellises of lichen. In 1897, the writer and naturalist John Muir had camped in Bonaventure for five days. He was penniless, and the cemetery drew him in. It was, Muir thought, a safe and quiet place to be. It was safe. It wasn't quiet.

"Many bald eagles roost among the trees along the side of the marsh," Muir wrote. "Their screams are heard every morning, joined with the noise of crows, and the songs of countless warblers, hidden deep in their dwellings of leafy bowers. Large flocks of butterflies, all kinds of happy insects, seem to be in a perfect fever of joy and sportive gladness. The whole place seems like a center of life. The dead do not reign there alone."

. . .

The number of missing servicemen from the wars in Iraq and Afghanistan was small, but the military had learned its lesson from the Vietnam conflict and was going to make sure no one was left behind. By the time the United States invaded Iraq in 2003, the military was quite clear that dogs could be useful for any number of things: bomb and land-mine detection, sentry duty, and enemy tracking and apprehension.

Kathy Holbert was one of the cadaver-dog handlers invited to apply to go to the Middle East as a contractor. Kathy runs a kennel in the mountains of Barbour County, West Virginia. Self-deprecating, self-sufficient, and humorous, she trains detection and patrol dogs, boards people's pets, and breeds a variety of working shepherds and Beaucerons, an ancient French herding breed. She occasionally throws a Malinois into the mix to keep things interesting.

Kathy had been in the military, first as a parachute rigger and then as a military-dog handler. That didn't go swimmingly. Her first dog, a "find 'em and bite 'em dog," appropriately named Dick, bit her at least a hundred times. "Actually, I was a terrible, terrible handler," she said. "My timing was awful. They used to use me to show handlers how not to do things."

That's hard to believe. Watching Kathy with both dogs and people makes the work seem simple, straightforward, and low-key. When Kathy got the call about going to the Middle East in June 2009, she was working her second cadaver dog, Strega, a sable German shepherd with an extra-long tail, big ears, and a witchy, mature intelligence. The decision about whether to go was oddly easy. Kathy and her entire family—her grandfathers, her father, her brother, and her husband, Danny, an electrician—had been in the military. She said yes. Then she thought, "You crazy woman, your dog's eight years old, and you're fifty, and you're going into hundred-degree weather. What the hell are you thinking?"

She did it anyway. Kathy remembered Vietnam. "They didn't put that much in trying to recover our boys at the time," she said. She took it personally. She and most of her family had served overseas. So Kathy started getting in shape, running, lifting weights, losing weight. She didn't want to be, as she said, "the missing link," the person who put soldiers in more danger than they already were.

Yet getting off the plane in September in Iraq felt like a body blow. "It's hard to describe the heat. It's like having a blow dryer in your face." A fetid blow dryer that smells like urine and blows sand at you. Kathy put booties on Strega, but they sometimes melted. The temperatures there average 110 in the summer. That's before one puts on heavy equipment and a flak jacket. Instead of trying to escape the heat, Kathy decided to embrace it. She stayed outside with Strega as much as she could. They both adjusted, and the experience made Kathy rethink what breeds and personalities of working dogs work best where. Strega, though a German shepherd, had a big, boxy nose but not a lot of huge muscle mass. She worked longer than some of the snipier-nosed breeds and seemed to do better in the heat than some of the big-muscled Labradors. She was a methodical worker, not too fast, not too slow, plenty of drive, but not flashy. Those qualities served her and Kathy well.

Greg Sanson, the personnel recovery advisor to the U.S. military in Iraq from 2009 to 2012, had a complex job: first to prevent kidnapping or abduction; then, once a contractor or soldier did go missing, to find him or her alive. If that failed, the next phase involved bringing in teams like Kathy and Strega. It was, he said, "an honor" to talk to me about the work Kathy did in the Middle East with Strega to help find the missing. "We don't quit looking for them," he said.

The work of looking was hard, both physically and emotionally. Improvised explosive devices (IEDs) shatter people. Kathy and Strega found themselves looking not for bodies but for small pieces of tissue.

Being able to slow a dog down when searching for someone blown up with an IED was terribly important. At first, Kathy said, just as at any explosion site, Strega didn't do perfectly. The scent of death was both everywhere and nowhere. Kathy understood that their job went beyond gathering enough DNA material to identify the victim. Kathy and Strega's job was different: to keep searching for anything and everything that could be found of someone.

Soon enough, with adjusted training, Strega understood the job. She started finding the little that remained.

14

Running on Water

"Believe me, my young friend, there is NOTHING—absolutely nothing—half so much worth doing as simply messing about in boats. Simply messing," he went on dreamily . . .

"Look ahead, Rat!" cried the Mole suddenly.

—Kenneth Grahame, *The Wind in the Willows,* 1908

David and I could hear Nancy Hook muttering in the kennels, getting the dogs watered and fed so we could leave. Our helping only slowed her down. "You son of a bitch," she told a massive pit bull mix. She kept her tone conversational. He had been biting people. Nancy would fix him. She specializes in dogs who bite without being ordered to.

Near the kennels, the bass boat sat on a rusty trailer, hitched to Nancy's pickup. Weeds had sprouted inside the boat. I plucked at them until Nancy arrived and told me to leave them alone. She was growing them on purpose, she said. Last night's rain had left an inch of murky water to water and fertilize the weeds. I offered to bail, but Nancy said a bit of water wouldn't sink us.

So off we drove, trailer bouncing, to Taylors Millpond. Two women, a man, a dog, a boat.

We were going to work on water.

Stories about dogs alerting on submerged bodies sounded vaguely apocryphal to me at first. But water is an ideal medium for transporting cadaver scent to a dog. Bodies seem to effervesce in water, like slow-motion Alka-Seltzer tablets. They are doing the same thing bodies on land do: decomposing and sending off gases. In the water, those gases bubble up to the surface and hit the air, then the dog's nose. Oils float up as well, providing a slick on top of the water that sends out additional scent.

Lakes or rivers can veil a body, though, even when searchers have the latest sonar equipment on hand. Often, especially in the Southeast, divers can't see their hands in front of their faces. Even when the water is relatively clear, diving and dragging don't always locate the body. A good water cadaver dog's nose can narrow the search substantially.

Solo and I had just one or two or twelve training issues. Nancy thought she could fix us.

Taylors Millpond is more lake-sized than pond-sized, more than a half mile long, created before the Civil War and now the site of bass tournaments. It's a few miles down the road from Nancy's farm. A general store faces the concrete boat ramp. A small group of men often sit on the concrete porch, chewing or smoking, cans of Pabst Blue Ribbon in hand. Today was no exception. I nodded, and they nodded back. The store owner shook her head vehemently as I tried to hand her the two-dollar boat-launch fee. She knew Nancy. She also knew we weren't there to drink beer and troll for largemouth bass.

We backed the trailer down the ramp and got the boat freed and floating in the duckweed without too much embarrassment and only minor slipping and sliding. David and Nancy planted themselves in the boat; Solo, who had already swum several laps across one end of the pond, leaped on board, spraying water over them. Solo loved boats. He thought it was fun to jump into them and even more fun to jump out of them. I pushed us off and leaped as well. The boat rocked woozily, and Solo climbed over us to the prow, a soggy figurehead. David fiddled with the trolling motor. He'd lowered it into the water, but it wouldn't start. He scowled. He was irritated. He likes things to work.

"It was free," Nancy reminded him. Like the trailer. Like the boat. Like Nancy working with us to train Solo on water cadaver. Acrid smoke oozed off a battery connector. I wondered aloud if batteries could explode. Nancy said no, but we knew she was lying. Bent over the dead motor, David noted that he could feel the boat moving. It was following the pull from Moccasin Creek. As we drifted closer to the edge of the overflow, I could see where the pond ended and water slid over the concrete edge in a fat silky ribbon and disappeared. I could hear the ribbon shatter ten feet below. I may have said something to David, because Nancy told me to stop giving him directions.

"Is she always like this?" she asked him. They smiled at each other. I shut up and took one of the paddles so I could save us. The motor sputtered to life and then settled down into trolling-motor Zen. The boat, finally under David's control, crawled away from the overflow and toward the center of the huge pond, dotted with floating islands of lily pads.

Water is not my element, though I like to look at it. My childhood swimming lessons were spent in a quiet panic. Water came up over my nose even in the shallowest end of the pool. I was short and skinny, with big bones that jutted out and no fat to help me float. I would sink, gulping chlorinated water; my long pigtails were ropes pulling me under while the instructor looked on, disappointed.

Water is Solo's element. He loves it and knows he has blanket permission—unless I specifically forbid it and sometimes even when I do—to dive into any available body of water or mud hole. Now he was singing a throaty paean that carried across the pond.

I could see his point. A great blue heron rose up out of the loblollies in primordial slow motion. Crappie, bullhead, bass, and catfish lurked beneath the lily pads and hunkered under the scrubby swamp roots that reached out into the alluvial floodplain. I couldn't see them, but a blue-gray-and-chestnut belted kingfisher knew they were there. She was perched on a snag, her outsize head cocked slightly. At our approach, she dropped and flew along the edge of the shore with a chittering rattle of irritation.

Though she was out of my league, I, too, knew how to fish. My grandfather, my father, and my brothers had taught me everything from worm-and-bobber fishing to fly-fishing. I knew how to creep up on a deep hole in a creek without casting a shadow or creating bank tremors that the brook trout could feel. In high school, I had made my own spinning rods, carefully layering thread to fix the guides onto the graphite poles. I stopped fishing thirty years ago. I wasn't patient, and I ended up feeling sad for the fish. I still like to eat them.

It was past time for me to return to the water. This time, I could let Solo find the fish. He was now seven years old. In the not-too-distant future, his increasing mental prowess would no longer be able to compensate for his slowly decreasing physical prowess. But as long as Solo had a good nose, water cadaver might extend his callout life.

That was why Nancy was pushing us out onto the water, toward certification. I'd turned down a healthy handful of water searches, and I hated saying no. One investigator swore that the victim had just walked straight out into the lake with his boots on, no way it was a criminal case. Could I please bring Solo? I was so sorry to say no. Mike Baker pointed out to me that all the investigators had to do was put on hip waders and walk straight out into the lake a few yards to find the victim themselves. Sometimes it is that simple. Sometimes the body floats. Mostly, it's more difficult. For people, that is. That's when dogs can help.

. . .

Dogs helped in a case in Tennessee a few years ago. The victim was last seen covering her boat on a long dock off an East Tennessee lake. Two dogs from Roy and Suzie Ferguson's Tennessee team were called in two weeks after her disappearance. The dogs both alerted on the dock, right where the victim was last seen. By that time, dragging equipment and underwater cameras were lying everywhere, complicating the scent picture. Searchers had worked nearly nonstop for two weeks around the dock area, with sonar, with deep-water cameras, with dragging, and with diving. Nothing. Investigators wondered whether the victim had left the dock. Or whether something nefarious had happened. A natural reaction when one doesn't have an answer.

The family didn't want to give up. They brought in an underwater construction crew with a deep-water robot from out of state. Roy and Suzie Ferguson came this time, along with the two other team members and the dogs who had originally alerted. Suzie brought her female German shepherd, Schatzie. Roy was the point man, observing the dogs from an opposite dock. It's always valuable on land searches if someone is there who knows how dogs work on land. Having a person who knows how they work on water is invaluable. Roy watched the dogs' alert patterns as they worked from boats and off other docks. Then he calibrated where the dogs alerted. The handlers and dogs did the same the next morning, when there was no wind. Roy reported the team's findings: The dogs had narrowed the area to a twenty-by-forty-foot oval. The crew put the little submersible robot, with its video camera and sonar, into the water at that spot. Using a joystick, they sent it down. The water was remarkably clear. In less than two minutes, the robot operators saw the victim, caught in the eye of the video camera. She was about thirty feet out from where she was last seen alive, covering her boat. That was in one dimension. She was 230 feet down, the equivalent of twenty stories beneath the lake's surface. University of Tennessee forensic anthropologist Bill Bass said that given the cool water, the depth, and the victim's fit build, she never would have floated. After falling, she probably floated down at an angle, away from the dock, flipping slowly like a leaf turning over and over as it drops from a tree. She managed, nonetheless, to send a final clear signal to the dogs.

· · ·

I entered upon the small enterprise of "learning" twelve or thirteen hundred miles of the great Mississippi River with the easy confidence of my time of life. If I had really known what I was about to require of my faculties, I should not have had the courage to begin.

—Mark Twain, *Life on the Mississippi,* 1883

Lisa Higgins knows water. She has the deserved reputation of being one of the top water trainers in the country. She also trains and deploys her own dogs; finding the time to do that on top of a hectic seminar schedule is, as all trainers know, a challenge.

Lisa was in the middle of a team training with Haylee when she got the call in July of 2011. Could she bring her dogs and come out to a Louisiana reservoir past Morganza, a reservoir that was part of a dam system for the Mississippi River? The system, challenged by record rains, wasn't perfect. The Army Corps had made the difficult decision to open the Morganza spillway and flood small towns downstream to relieve pressure on the levees in Baton Rouge and New Orleans. The reservoir was still high in early July when a family of four went out to fish. It was an area the family knew well, but the water level created a churning boil at one spot. Their boat stalled on a log, and the boil pulled it in. The boat capsized. The father managed to boost his wife and one of the boys to safety on the spillway wall. He got his second son over to the wall, and his bruised and injured wife was able to pull the boy to safety. That last effort was too much for the father.

"He saved his whole family. He was too tired to help himself," Lisa said. She and another team from Jefferson Parish got the call; Lisa was with Haylee in a spot where it wasn't easy to get Haylee home first. Law enforcement on the scene said sure, bring her along. Haylee, being Haylee, was thrilled. Lisa, being Lisa, figured that, carefully handled, it was another opportunity to school Haylee away from home. Law enforcement on the scene were "exceptionally wonderful to her," Lisa said.

That first day was long. When it's hot and windless, it can be even hotter inside a boat. Lisa was working both her dogs, Dixee and Maggie, along with the Jefferson Parish team, with their dogs. Because of the heat, handlers were working their dogs twenty minutes each. They gridded, worked the dogs, rested and cooled them, then worked them again.

Toward the end of the first day, "I noticed that Maggie thought she had an area," Lisa said. They marked that and one other area of interest for local law enforcement and divers. They didn't have time to narrow it down more that day.

Authorities called Lisa back on July 4 to keep looking. Law enforcement on the scene were a bit disappointed Haylee couldn't come that day; she had been good company. Lisa started in the area where little Maggie, her seasoned Australian shepherd, had reacted a couple of days before. At the end of thirty minutes, Maggie was panting. Lisa put her up in the truck to cool. She talked with the people on the scene. Dixee, her intense Malinois-German shepherd mix from Kathy Holbert's kennel, had never had a water recovery. On the other hand, she wasn't exhausted, like Maggie.

Dixee went out. Dixee alerted. Law enforcement recovered the victim there that night. He was in 129 feet of water, more than 200 yards from where an eyewitness had seen him go down. "Bodies can travel much further than that in water," Lisa said.

It's not just the horizontal distance. Water searches have a three-dimensionality that can make it difficult, if not impossible, to find bodies, even using the latest technology and divers. Depending on electronics is bound to disappoint. Side sonar can help, but if you've got a search area as large as a reservoir, a body can be a needle in a haystack of objects at the bottom: boulders, logs, bushes, snags. Or a body can be suspended between the bottom and the surface because of temperature gradients and currents. People who do recoveries in cold water say that if the person went in alive, he can curl up in a fetal position and end up on the bottom, camouflaged as a large rock. The best side-sonar scanner in the world won't help distinguish that rounded shape from the others.

There are drownings with fluid in the lungs and cases when the person is dead before going into the water, whether by accident or murder. Victims who are dead before entering the water, or get only a bit of water in their lungs, tend not to sink. Each factor affects the disposition of the body and whether it floats or sinks; it also matters whether and what the person has eaten before going into the water, and whether the person was weighted down with concrete, wrapped in a tarp, or wearing hip waders.

Then there's alcohol. Whether or not it played a role in the first place, and it often does, it definitely plays a role once someone is dead. Beer bloats. "If there's beer on board, refloat time could be a lot quicker," Lisa Higgins told handlers. In Louisiana, Lisa also considers whether there's a Creole influence in the mix: Red beans and rice speed float time.

Serious water cadaver-dog handlers carry around a "body float" chart; many of them know it by heart. It provides information on water temperature and estimated days-to-float time, along with variables like body composition.

Water cadaver training works just like land cadaver—finding the strongest source of scent. Then it gets hard. At first it's canine torture, like tying a dog's feet together and then commanding him, "Go find!"

The dog doesn't get to run around and climb over obstacles to find scent. Instead, he's trapped in the boat, which has become his legs, and he's forced to communicate with scent illiterates to make those legs move anywhere.

"We're taking a lot of things away from them," Lisa Higgins warned handlers during one lecture. "We're taking away their ability to run the scent cone with their own four legs. Now they are dependent on us."

Water search was taking patience away from me. Solo and I were not the perfect pupils. Neither of us fully understood the concept of what we were doing on water. Training a couple of times a week on water wasn't possible. We'd already been out in a boat with Nancy a number of times. It was trigonometry to me, trying to understand the currents, the wind, the scent cone, the boat chugging in a zigzag across the water. For Solo, water work was hyperbolic geometry. It takes patience and nerve strength on the dog's part. Solo has a great deal of drive and good nerve strength, but less patience.

Solo also needed to learn to modulate, so that if we were on an actual search, the buoys that searchers threw down each time he alerted would define a small area for dive teams, not an entire lake of dog screams. Sending a diver into murky water with snags or dangerous current, based only on an insufficiently trained dog's alert, isn't just a waste of time. It's endangering living people to recover a dead one. The first three times we went out to train on water, Solo screamed incoherently, even when we were a hundred yards or farther away from the scent source, lunging in frustration for the reward toy, which I had stupidly placed in the breast pocket of my life preserver.

Solo was in scent, just not the strongest scent. I threw up my hands, Nancy laughed and shook her head, explained the principles again, and around the lake we went once more. Nancy wanted him quieter, and she wanted him closer to the source. I agreed.

Dogs who have worked on land have to learn a whole range of new behaviors in a boat, where a dog needs to give increasingly strong cues to the handler as the scent gets stronger. Solo needed to learn to use what little physical freedom he had to tell me and the boat driver to follow his lead, by turning this way and that, or by going from the front of the boat to the back. He needed to communicate with the captain that the boat was moving into or away from the perfume rising from a body far beneath the water's surface. Solo, working on land, is usually able to run where the scent is strongest and lie there. He can work out problems on his own timetable, not that of a moving boat. If I know he is honing in on something, I can slow down a bit and let him work. Water is a different gig. Instead of just moving himself closer and closer to the strongest emanation of scent, he needed to tell the boat driver exactly what he wanted. His bossy attitude would come in handy—if he could learn to give clear directions.

Andy Rebmann wrote much of the book on cadaver dogs, but Marcia Koenig wrote the chapter on water cadaver. It was only after studying the illustrations in that chapter, watching Solo on water, and watching a bunch of experienced water cadaver dogs that I started to fully understand the fundamental differences that dogs—and their handlers—face in water work. And how dangerous a "three-dimensional problem" can be for everyone when flood stages, boiling currents, and logjams are in the mix.

Marcia researched and re-created an extraordinary search and recovery using dogs on a flooded Ohio River. Her hypothesis is that dogs alert where water breaks around an object, where scent gets strongest, and then suddenly disappears. The handlers on the Ohio River search all noted the same phenomena. Nikki, a German shepherd, had an entirely new reaction: Her bottom jaw vibrated as she worked scent. She was still nearly a mile and a half downstream from the victim. As the boat got closer and closer to the body, Nikki gulped water, spat it back out, clawed, tried to jump into the water. When the boat crossed the invisible scent line, just upstream of where the body was located, she visibly relaxed. She smelled nothing but fresh, unscented water. "Nikki went completely limp for just an instant." Game over. Scent, scent, more scent . . . no scent.

Her handler dropped a buoy on that line. Marcia wrote: "She remembers thinking, 'What have I done?'" Nikki's handler had never seen her react that way. Yet as one of the dog handlers said, for the dogs, "it was like stepping from one room to another."

That threshold was where the victim was located, trapped under logs.

Water work, even more than land work, depends on being able to know and read your dog, and having someone along who can watch closely. The dog's cues may be much different and run the gamut between subtle and dramatically obvious.

On land, Solo uses his whole body, including his large tail, and I get to watch him from a distance. He has a much smaller stage to work on when he's on a boat; he's working in close-up. We'll need to work with a whole series of clues, escalating signals that can help the handler and the boat operator work in tandem. Licking lips, lapping water, throwing his head: It's a new vocabulary. Because water work is so different, Solo might also need to find a new way of alerting, so there's no question in anyone's mind that there's something there. Biting at the water, yowling, who knows? I don't fully know what Solo's alert will be on water. We're still working on it. I hope we get there.

Detective Art Wolff's gorgeous Belgian Malinois, Radimir, whose name means happiness and peace in Russian, digs at the bottom of the boat. If he could just dig through, he knows he could get to the body lying in the water beneath. He has gone all over Tennessee on water recoveries.

On one of the most emotionally difficult cases that Canadian trainer and handler Kevin George ever worked, on a flooded river in Calgary that claims victims each year, his Belgian Malinois started barking, leaning over the side of the Zodiac, snapping at the water. In his excitement, he snapped at the side of the inflatable boat. Thankfully, he didn't deflate it. Sadly, the river was too high and filled with unstable logs to recover the victim safely.

Nancy liked what her big German shepherd, Indy, used to do: When he couldn't stand it anymore, his big question mark of a body would become more and more unbalanced over the water until he would tumble in where the scent was strongest.

Then there's an even more direct method. As Lisa Higgins told a group of handlers, "A dog jumping out of the boat and swimming in a circle over the source? That is an indication you just can't miss."

Lisa does worry about alligators where she works, so she doesn't encourage Louisiana handlers to let their dogs jump out of the boat to alert. On one search, she counted thirty alligators. Interestingly, they hadn't touched the victim Lisa was looking for. Nonetheless, for training, she chooses her spots carefully.

"I go to places I know to be safe, because a 'gator may not attack me, but it will have my dog for lunch."

. . .

Friends, if we retreat to the hollow ships, and yield this body to the horse-taming Trojans, who'll drag it to their city and gain the glory, that would be small fame indeed for us: better the black earth swallow us instantly where we stand.

—Homer, the *Iliad*

Since their arrival in Iraq in September 2009, Kathy Holbert and Strega had started to adjust to the oppressive heat, the wind, the sand—and the danger.

In November, just two months after their arrival, the military sent them into Afghanistan, fifteen hundred miles to the north. As they flew in, Kathy could see the country's wild beauty below: its glorious tan and ocher mountains, its winding rivers, its stunning farms and orchards, which were finally starting to recover from being razed and bombed by the Soviets during their nine-year occupation fighting the mujahideen. Kathy knew the beauty would be offset by what she and Strega would face on landing: a harsher and even more dangerous environment than Iraq.

"Afghanistan was different," she said flatly. "They try to kill the dogs."

Military K9s and handlers get targeted there much more purposefully than in Iraq. Afghan fighters know how demoralizing it is when the U.S. military loses a handler or K9. Those are the teams that go out first. The dog sweeps a hundred feet or more in front of his handler, searching for the scent of IEDs. Both dog and handler are ahead of the troops. Some of the IEDs have wires running into nearby ditches where someone waits to detonate them. First out. First to die. Kathy keeps an increasingly long list of dogs and handlers killed.

It's not only the Taliban and their sympathizers who pose a danger. In Afghanistan, Strega was just another occupier. Children, mostly girls, followed them after they landed. They were beautiful, Kathy said. She turned to greet them and hand them candy. They countered by throwing rocks at Strega. "They tagged her pretty good."

Soon, Kathy and Strega were in a convoy, heading farther north to a search in the Murghab River, near the border of Turkmenistan— a six-hundred-mile-long river that flows north and disappears in the sands of the Karakum Desert. Two paratroopers had drowned. It was a classic tragedy—a crate of supplies dropped out of an aircraft by parachute that ended up in the river instead of onshore. The first paratrooper was in full battle dress. He waded out on the shelf of the river, where the water was calm and shallow, and grabbed the supply crate as it floated by. He must have been pulled hard off the shelf's rim by the palette's weight just as it careened into the fast, deep water. It would have been like grabbing a floating boat that suddenly revved up its engine and took off. Seeing his fellow paratrooper in trouble, another trooper went after him. And like that, he was gone, too.

The Taliban, downriver from the site, claimed they had both men's bodies. Thus began a massive recovery operation for the two deceased men. A team of British divers flew in and started searching. After a week, they found the paratrooper who had gone in after his friend. His body was around the bend in the river, well downstream. The recovery operation came at great cost. Downstream was filled with Taliban fighters. The Associated Press reported that eight Afghans were killed—four soldiers, three policemen, and an interpreter. Seventeen other Afghan soldiers and five American soldiers were wounded.

And no second body surfaced. That was when Greg Sanson, then the personnel recovery advisor for the U.S. forces in Iraq, got the call asking for a human-remains dog team. He sent Kathy and Strega.

Water was a familiar search area for Kathy. That was where she got her start with her previous dog, Mangus. Kathy was a sheriff volunteer with a narcotics patrol dog when she was called to the scene of a drowning one summer. She sat with the family whose young son had drowned. They couldn't find the body. The family knew, at some level, that he was dead. Then again, they didn't.

"This went on for five days," Kathy said. "I don't think that people understand what families go through when they don't know. One of the things I learned is you don't ever, ever say the word 'closure' to that family. There's no such thing."

Those wretched days started her thinking. Could a dog have helped? A state trooper told her about Charm Gentry, a cadaver-dog handler in another part of the state. "I contacted her, and it just so happened they were getting ready to do an Andy Rebmann seminar."

Kathy certified Mangus by the end of the seminar. The week after that, he helped pinpoint a drowning victim. Mangus ultimately helped recover twenty-seven people. He made it look easy. He went to the front of the boat, lay down, and put his head over the bow, close to the water. Just like that. A natural.

Strega, though, had her own ideas about how she wanted to run a water search. She kept wanting to leap into the water. "She was a very stubborn dog," Kathy said with pride. The two of them finally worked out a system that made them both reasonably happy.

And then there they were. In Afghanistan, to work on water.

The terrain was steep, with willows along the edge of the river. The river looked placid on its east side, where it was shallow and calm enough to reflect a hint of blue sky and a slightly muddy color, with the sandy shelf just visible underneath. Farther out, the water turned white with froth, a churning, greenish gray. Kathy suggested she and Strega start searching at the place where the men fell in. They said they were sure the body had moved on, beyond the bend of the river. I can picture experienced, low-key Kathy quietly saying, "No, you need to go back to the point of entry."

Kathy started Strega at the point of entry at ten A.M., on the east bank of the river, where it was warmer in midmorning, with current to move the scent around. It wasn't just the river they had to deal with. A sheep-and-goat guard dog—powerful and leggy—headed straight for them. He stopped at the last moment, retreating to a ridge just above the river. He looked to weigh at least 125 pounds. "He watched the whole time we worked."

Strega worked to the bend that the military suspected the paratrooper's body had gone around. Then she tried to get into the river. She wasn't interested in going around the bend. She moved back upriver and worked the rapids. She alerted on the shore, across from a churning area yards below where the men were last seen. It was her simple alert. A sit. It told Kathy everything she needed to know, although she suspected the divers, who had never worked with a dog, were probably hoping for more "yippie-yi-yo-ki-yay." Strega was not a dog who did backflips. Her sit, nonetheless, started to recalibrate everyone's thinking about where the paratrooper's body might be.

The next day was a Sunday. Not a day of rest for anyone. The humidity and wind were better, and Strega told everyone once again that she was pretty darned sure the paratrooper had never made it past the bend in the river. She kept working upriver, closer and closer to the boil.

The divers rigged a more complex system, a high line over the river, and put Kathy and Strega on a rubber raft. Back and forth. Strega alerted right over the rapids, midway across the river. Kathy and the divers looked down to see a deep undercut and a hydraulic boil.

Imagine a washing machine in its highest spin mode with lots of water. It keeps tossing clothes around but never lets them escape. That's a hydraulic boil, more familiarly known as "a drowning machine." The British divers tried to search it, but the currents were too strong. They did manage to snag the parachute and cargo box. Both had been trapped in the boil.

They had to trust Strega. A number of people had already died. And while the military talked about a couple of different options—including putting some blasts into the river—they decided to let nature take its course.

"K9 assistance was ended," the report read. Strega had been utterly consistent over three days on the same area of the river. You shouldn't ask a dog to tell you the same thing over and over, seeking reassurance, like you do with people.

The military notified the villagers that there would be a reward if they watched the river. Kathy, who knew her float chart by heart, provided the military with her estimates of when villagers might see something, given the days since the paratrooper disappeared, what he was wearing, and the temperature of the water. The hydraulic boil made it hard to know what might happen.

Nearly two weeks after the search was suspended, villagers contacted the military. It was late November. The paratrooper's body had surfaced right in the area where Strega had alerted. Strega, a dog trained to find human remains, had probably helped save lives. The Taliban didn't get the body. No one else died on that mission. No one triumphed, either.

15

The Perfect Tool

In this business it takes time to be really good, and by that time you're obsolete.

—Cher

Just as I got a handle on dogs' noses, appreciating them for their precision, practicality, and adaptability, science leaped ahead of me. Something superior to the primitive dog snout was around the bend. I was a scent Luddite, hunkered down and defensive of the working-dog class at the moment the biotechnology revolution was launching something slick and superior.

Time magazine ran a piece about the demise of working dogs one week before Joan Andreasen-Webb told me that Vita was pregnant with Solo. Before Solo was more than a clump of rapidly reproducing canine cells, he and his nose were passé.

"Memo to man's best friend," *Time* told magazine-reading dogs in January 2004. "In a few years, you may be relieved of your police drug-sniffing duties, thanks to a pair of Georgia Tech scientists." Scientists "have developed a handheld electronic nose that detects the presence of cocaine and other narcotics better than your cold, wet snout ever could." The magazine overestimated dogs' print literacy while not giving them nearly enough credit for nose literacy.

This biotech moment wasn't the first time I'd felt like an apprentice on the cusp of entering into a brilliant new career just as pink slips were getting handed out. My career timing had always been lousy. In 1982, the same year I started my first newspaper job in the foggy San Joaquin Valley, Gannett launched *USA Today* in full color, with its nuggets of fun-to-consume news. Publisher Allen H. Neuharth called the style of reporting and writing the "journalism of hope." Headlines emphasized the positive. When a charter plane crashed in Málaga, Spain, the headline read: MIRACLE: 327 SURVIVE, 55 DIE. But newspapers were already ailing, victims of executive bean counters, their own stockholders, and, soon, the burgeoning Internet.

Thinking quickly, I transitioned from newspapers about a decade before massive layoffs began. This time I trained in an industry I knew would last. Higher education had been around since medieval times. Knowledge would never go out of fashion. I became a tenured university professor about a decade before tenure started dying and the sun started setting on affordable public universities.

Now the working-dog nose was in decline as well. Scientists, engineers, and chemists, with the media as their cheerleaders, were informing me that my new avocation was well on its way to obsolescence.

Dog substitutes—biomimetic replacements—weren't just hot; they were blowing dogs out of the scent pool entirely. Once again, I had arrived at the tail end of an era.

Researchers know—thanks to the public relations officers attached to their start-ups and universities—that they need to market their fake noses using familiar and fuzzy terms. The fake noses might not be surrounded by furry muzzles, but the image needs to be there: FIDO, RealNose, the E-Dog. Sometimes names backfire. Dog-on-a-chip was not a good choice.

All of these artificial noses—whether for bomb detection, drug detection, land-mine detection, or human-decomposition detection—have several things in common, according to the media and the grant applications. They won't shed and won't bite. They won't get tired or overheated. They will detect parts per trillion of anything. They will put sniffer dogs out of business. Any day now.

Like the cure for cancer and intelligent artificial intelligence, the ultimate mechanical nose is on the job sniffing, just around the bend of the next news or grant cycle. Every time there's a bit of engineered-nose news out there, the media shifts into high gear. People love a couple of things: way-cool technology and dogs. Perhaps cool technology is hotter than the ever-faithful canine, but we know dogs will be there waiting for us even after we've hooked up with the latest gadgets.

Just a year and a half after *Time* magazine informed dogs that they were working on borrowed time, I started Solo's serious training with Nancy Hook. I was able to read and understand his behavior changes. I knew what happened to scent in heat, in wind, in rain, and water. I was Dr. Dog Science.

That same summer, Glen Rains at the University of Georgia and his colleagues filed their patent for "The Wasp Hound," an eight-inch portable tube filled with starved parasitic wasps, a fan, a video camera, and a computer. When the target scent wafted into the chamber—whether it was the scent of human remains or of a bomb—the trained wasps would congregate at a pinhole, hoping for their food reward.

The video camera watched the wasps and signaled their behavior change to a computer. The patent application called dogs "subjective and costly." Wasps were "more sensitive, programmable, portable, and cryptic." That last adjective was critical in a post-9/11 world. On that point alone, the small, silent wasps in the tube triumphed. From the day he was born, Solo was never, ever cryptic.

At least I could understand the concept of the Wasp Hound as I looked at the application's illustrations. I couldn't say the same about the dog-on-a-chip, although its inventors called it "an elegant fusion of biotechnology and microelectronics." However, Solo wasn't any more elegant than he was cryptic. He did have more in common with some components of the Wasp Hound than with many of the bioengineered noses. He and the wasps might not be same species, but they both belonged to the kingdom of Animalia.

. . .

Mechanical noses, mixtures of mechanisms and organisms, or their newest iteration, genetically engineered cells, have been a Holy Grail for applied scientists. Some researchers continue debating the physics of scent while others labor to understand what dogs are alerting on. A third group has dismissed the dog nose altogether and is working on its replacement. Though the dog's nose might be a black box and a real challenge to reverse-engineer, no scientific law says you can't try to build a better black box before you understand what's inside the original.

The first artificial nose became available in 1982, the same year *USA Today* was launched. The "Warwick nose" had a sensor made of tin oxide and is still produced for commercial uses. In 1988, *Computer Business Review* noted that drug kingpins could use masking odors, "leaving poor pooches puzzled," but the Warwick nose could do better. The only thing needed, the *Review* urged, was more development money; otherwise, Japan might win the electronic-nose race.

We've lived with the term "electronic nose" for nearly three decades, and despite some very different mash-ups using gas chromatography, grids of nanoparticles, or polymer sensors, any artificial nose needs to do three things: take in a vapor, show it on an array, and identify it to the machine's operator. The dog's nose does something similar: It takes in a smell and sends a signal to the dog's brain. The dog sits: I smell dope. The handler sees the dog sitting and knows she's just smelled dope.

One cannot patent the dog. So "a truly biomimetic olfactory microsystem"—that is, a fake nose—became a goal for bioengineers. Despite its lack of a mammalian reproductive system, the fake nose is a gift that keeps on giving. In early theoretical stages, it helps get papers published. In experimental stages, it helps buy lab space and equipment and pay for postdoctoral lab workers. If the nose ever goes into production, even more people get involved. Most of the current fake noses are expensive but not too expensive, especially in these days of downsizing public infrastructure. Law enforcement, the military, health care, and other industries, such as food production facilities, will buy it. Developers admit that the fake nose will require some maintenance and training. That admission just brings in additional funding.

As each new nose is trotted out to the media and public, the same claims are made: This nose is the nose to end all noses. It will keep working in heat and cold, never false alert, never get tired, never require dog chow. It won't have to retire at the age of eight or nine with hip dysplasia.

Mechanical noses have flourished under scientific competition. But like a bad reality show, *So You Think You Can Make a Better Nose*, the competition can be a polyphony, or even a cacophony, of competing tunes and dueling harmonies: Anything a dog can do, a machine or a mash-up can do better. Your fake nose is good, my fake nose is better. Anything a Georgia Tech chemical engineer can do, an MIT researcher can do better. "There's no further improvement in the sensor part you can get," an MIT chemical engineer told *Wired* magazine about his e-nose creation in 2010. "It's the last word in sensors."

It may be the latest word in sensors. I doubt it's the last. Call me unsentimental, but I can't decide which I prefer: having an airport security employee thrust an electronic nose into my crotch, or having a TSA handler do the same thing with a bomb dog. If my civil rights are going to be violated every time I fly, I do prefer that the search be effective and low on the false positives. And that I not get cancer or get bitten.

. . .

A couple of decades before the term "electronic nose" came into being, army researchers, much like those at Southwest Research Institute, started thinking about replacing the dog, a sometimes unwieldy biological system, with a better system.

The experience of using sentry dogs in Vietnam in the early 1960s made for some imaginative leaps of faith, and while some army researchers, like Nick Montanarelli, stuck with the tried and true and played around the edges by trying different breeds of dogs for scent detection, other researchers at the U.S. Army Limited War Laboratory went further. Dogs' noses were great for detecting the enemy, and dogs' barks were great for arousing sleeping soldiers. What about partly mechanizing and miniaturizing that system?

So in 1965, army researchers started playing with "insect ambush detectors," a reasonable riff on the reality that blood-loving insects were ideal candidates for the "bio" part of the machine: ticks, mosquitoes, bed bugs, giant conenose bugs. All those insects use the warm breath of vertebrates to find their next meal. Researchers rigged up a plastic tube with a bellows, a microphone, and a sanded piano wire that would twang when the bugs' feet started a frenetic dance because they sensed a nearby meal. Ta. Da. An "ambush detector," ready for a wild rumpus when it smelled the breath of enemy soldiers.

Early work narrowed down the best bugs for the job. When fleas sensed humans, they started jumping violently. They wouldn't settle down. They had to be fed too often. Bed bugs, like fleas, got overly excited about the prospect of a hearty meal. Ticks were an early possibility. They loved the smell of human breath and could move quickly, though they didn't jump. I've watched them crawling up my arm, so I know. They have soft feet. That's why you think you can feel something crawling on you, but it's just a vague sense, and you forget about it. The next time you think about the tick is when someone tells you one is attached to the back of your neck. This is advantageous for ticks but not for army researchers, even with a microphone inside the detector. The researchers tried tying weights to their tiny tick feet, but they still were too silent, even wearing clodhoppers.

Conenose bugs—commonly known as kissing bugs because they most love the fleshy part of your face or lips for their bloodsucking—were just right. Until the researchers tested them. The bug machine performed miserably in the 1966 Panama Canal field test, the army report noted. The false positives were through the roof. Once the conenose bugs got amped up, they were as bad as the bed bugs: They refused to settle down. Their feet just kept noisily sending out signals. Food! Food! Researchers realized it wasn't just food. Motion excited the bugs. Wind excited them. Pretty much everything excited them. It was like a kids' sleepover.

In what would become a template excuse for not finding something as good as the dog's nose, the final report said that using blood-loving bugs as ambush detectors remained "technically feasible." That term is a big ol' hairy clue to recognize a failed experiment.

In any case, bugs are passé. In the past five years, the entities that have truly turned on tech reporters aren't the hybridized machines with animal names, like the Wasp Hound, but plants. Bomb-detecting ferns, for example, created a minor media stampede. "They turn completely white when they sense something nefarious around them," a FOX News reporter said. Since we are used to seeing big leafy plants in malls and in fern bars, I worry that we'll hardly notice them turning from a robust green to a sickly white if someone plants a bomb.

"We actually modify the seed," University of Colorado biologist June Medford told FOX News, "and then it's a trait that is stable and stays with it forever. It's very empowering because it will tell you that there's an explosive around: 'Get the security guys here!' "

Sounds great. So how long does it take that fern to turn white and tell the security guys? While Medford says that her studies show the plants have "detection abilities similar to or better than those of dogs," it takes the ferns hours to change color. "Work is under way to reduce that to a few minutes," the report promised.

We don't yet have bioengineered green dogs who turn white when they smell a bomb. The old-fashioned ones are pretty fast at telling us something is amiss. Faster than hours. Or a few minutes. More like a second. When there's a bomb, speed can matter.

That's the problem with the vast majority of the dog replacements, from fern to machine to all the things in between: They don't have the skills that dogs have in one furry package. Researchers realized that in the 1970s when they tried a variety of species to replace dogs; they keep learning that lesson over and over. Dogs are adaptable, mobile, sensitive. They use complex cognitive judgments to avoid lots of false positives. They can do several things at once: sniff, raise an alarm, bite if necessary, act as a deterrent. They interact with people more than a fern does. They're more fun to have around than angry bees. I know this, having been stung a few times tending to our beehive.

Best of all, dogs are comparatively inexpensive. The argument that it takes money and time to train a dog is countered by the argument that training technicians on machines can be just as expensive. Machines don't run themselves any more than dogs do. Machines break down, they need calibrating constantly, and they can be more temperamental about weather than dogs.

"When I started doing work in this area twenty years ago, I originally thought we would be able to make a machine that could replicate a dog," said analytical chemist Ken Furton. "But it's not going to happen in my lifetime. We are not going to replicate what a dog can do."

The Pentagon came to the same conclusion in late 2010 in Afghanistan and Iraq. It shut down a huge program that had spent $17 billion in employees and technologies and failed to make a dent in the problem of IEDs. After five years, hundreds of projects, and a "blizzard of cash" paid to the country's biggest defense contractors, reported the Center for Public Integrity, one system rose above the others: dogs and their handlers, along with observant people: "The most effective IED detectors today . . . don't hum, whir, shoot, scan, or fly. They talk. And they bark," Peter Cary and Nancy Youssef wrote.

The rate of finding IEDs with other technologies stood "stubbornly" at just 50 percent, Lieutenant General Michael Oates told *National Defense* magazine. Handlers and dogs found IEDs at the rate of 80 percent. The best bomb detectors, Oates said, are dogs working with handlers, local informants, and the trained soldier's eye.

"That combo presents the best detection system we currently have," Oates said.

A couple of years ago, I happened to be standing next to one of the best detection systems we currently have. I was at a K9 training near a military base in North Carolina. A special forces handler and his military working dog were taking a break from classified work in the Middle East. The Belgian Malinois was lying quietly beside his handler with his back legs tucked under his haunches and his front claws dug slightly into the ground. His eyes weren't fixed on his handler; instead, he was looking out. The dog's eyes never stopped scanning the big field of mowed grass in front of them. The handler looked down at his dog. It was, he said, like having a formidable extension, two and sometimes three hundred feet in front of you, keeping track, watching out for you, making sure you weren't attacked. Or blown up. The handler said he couldn't count the number of times his dog had saved his life.

. . .

Not only can turkey vultures detect a dead mouse from more than 1,000 meters (3,300 feet) away, but they have a major advantage over sniffer dogs—they can fly, removing the challenge of difficult terrain.

—*Der Spiegel,* 2011

Bomb detection isn't the only arena where machines and their mash-ups don't appear to be making much progress. The body-location business isn't seeing a great deal of electronic nose success, either. One group of German forensic scientists said in their 2010 conclusion about using machines to locate human bodies (though they used dead rabbits in their experiment): "In principle, an electronic nose based on the sensors applied in this study can be used to find decomposing human bodies in terrain. However, for design and development of a practically applicable device, sampling and measurement procedures have to be optimized."

In principle, with optimized hooves, pigs can fly.

Between the Germans with their rabbits and the Georgians with their wasps, I was starting to feel better about the dog's place in the body-location universe. Good dogs seem to move through a kind of complex decision tree on difficult searches: "This, not that," "Up, not down," and "That thing doesn't belong here, but it's not the thing I'm looking for." Although dogs aren't perfect, they adapt to a variety of search conditions.

Nonetheless, challenges to the cadaver dog's primacy are constant. Perhaps it wouldn't be a machine but a bird that would do the trick? No researcher ever wants to give up on the possibility that other species are better at scent detection than dogs. That's why I was worried when I first heard about the turkey vulture experiments in Germany. Although the studies are limited, turkey vultures appear to have the most advanced sense of smell of any raptor. The idea to use them to find human bodies was ingenious. A search dog with wings and a better nose. The Hanover, Germany, police commissioner, Rainer Herrmann, told fascinated reporters that vultures, with their ability to fly high and cover kilometers of forest, might have a major advantage over sniffer dogs. More than forty agencies from Switzerland, Austria, and Germany expressed interest.

It wasn't the first time people had thought about using vultures as sniffer animals. As early as the 1930s, the famous ornithologist Kenneth Stager reported, oil industry engineers in Texas had added ethyl mercaptan—a chemical that smells like carrion—to pipelines, then watched turkey vultures to see where the leaks were.

For this twenty-first-century version of search vultures, a few small hurdles had to be overcome: The vultures needed to be properly trained, equipped with GPS locators, encouraged to find only dead people, not dead animals—and convinced not to swallow the victim and destroy evidence before the police arrived. Vulture trainer German Alonso told reporters that police probably could arrive in their cruisers before too much important information went down the vultures' gullets, as the birds tend to peck rather than devour. I studied the photos of Alonso with his search vulture in training, Sherlock, perched on his arm. Though I was a sucker for a good-looking German shepherd, I figured I could get over the bird's bald, wrinkled red head and huge Roman beak. Solo has a Roman nose. But I couldn't help thinking about how vultures like to throw up corrosive vomit and how they pee down their own legs to keep bacteria at bay.

Sherlock didn't like to fly when he was searching for his training material. Instead, he waddled around like a duck. He was so anxious and antisocial, Alonso said, that given the command to search, he would hide in the woods or bolt. Miss Marple and Columbo, two younger vultures brought in to assist Sherlock and make him feel as though he were part of a big vulture family, fought constantly. None of the vultures seemed to give a fig about the difference between animal carcasses and human cadaver.

The Social Democratic Party—the opposition party—suggested that the state government start an international training center for search vultures. By the time the laughter faded, so had the project. "The vultures are not currently available to journalists," bird spokesman Stefan Freundlieb told *Der Spiegel* reporter Michael Fröhlingsdorf.

For the time being, neither dogs nor people nor vultures nor machines can do it alone. This is especially the case with clandestine burials. A decade-long FBI study on clandestine graves points to the problems of detection: The average age of the burial at the time of the search was four to six years. The bodies were typically off the beaten path, away from traveled roads and paths, surrounded by heavy brush, and buried up to two and a half feet deep.

Unless a murderer confesses and provides a detailed and, more importantly, correct map, looking for a grave takes an enormous number of investigator hours. Then there's investigator bias, cadaver-dog handler bias, forensic archaeologist bias. You can end up with holes dug all over creation. It's exhausting, discouraging work.

"Personally, I think dogs are invaluable for this type of work, but I don't think they should be used alone," Arpad Vass said. "I'm very cautious about digging a hole where a dog alerts."

That's not because Arpad doesn't trust good cadaver dogs. He's a fan, and the dogs are doing their best. But scent moves, chemical plumes move, decomposition moves. A victim's remains can be hundreds of feet away from where the dog alerts.

"You need a backup plan," Arpad said. Several, preferably. Ground-penetrating radar can help, but GPR throws false positives as well and can't be used in all terrains. Also helpful are a hydrogeologist, a magnometer, a metal detector, and great investigators who manage to set aside their preconceptions.

Last, a machine that can recognize the four hundred or so volatile compounds we vent as we head back to dust. That's why the National Institute of Justice decided to support Arpad and his colleagues' work to create a machine that could measure "odor mortis."

Enter Arpad's LABRADOR. That's an acronym for Lightweight Analyzer for Buried Remains and Decomposition Odor Recognition.

It should be in production this year. It looks like a metal detector. It's not meant to replace anything. It's meant to complement the whole kit and caboodle—investigators, ground-penetrating radar, geologists, forensic archaeologists, hydrologists, magnometers, and cadaver dogs and their handlers.

LABRADOR's early promotional literature, probably because Arpad helped author it, was modest: "The sensitivity of the instrument does not yet compare with that of a canine's nose." Arpad is human, though. He couldn't resist one additional piece of furry marketing to his prototype beside the acronym: The silhouette of a square-muzzled hunting dog once graced the early machine's instrument panel. Sadly, the company producing Arpad's machine decided to remove both the silhouette and the original name. I doubt it's because they prefer German shepherds.

16

Grave Work

The holiness of nature is ever a lofty contemplation; and it is well amidst the quiet wildwood and beneath the forest-shades, to be reminded some-times of death and of the grave. . . .

—Nehemiah Cleaveland, *Green-wood Illustrated,* 1847

When I walk in the Piedmont woods with Solo, on a search or for pleasure, I wonder when he tarries on a particular scent whether it's more than squirrel pee or the ancient track of a pit bull.

As a Yankee, I sometimes paint the Southeast's history with the crude black-and-white brushstrokes of slavery, but these walks remind me that the South's dead go back thousands of years. Once I start to think about who might lie beneath the forest floor, my perspective broadens and deepens.

Historic human remains, as they're known in the cadaver-dog world, can be a distraction during missing persons searches. One long day's work around an abandoned plantation, in a case that was barely cold, ended with Solo sniffing and working the downhill side of a slave cabin foundation with great interest but no final alert. I watched and rejected his interest as insignificant to the search we'd been called to do. As I drove home, exhausted, I realized how much birthing, living—and perhaps dying—must have occurred in that dirt-floored cabin.

Kentucky coroner Barbara Weakley-Jones, who founded and directed the Kentucky State Cadaver Dog Program when she was with the medical examiner's office, said that she doesn't like to train her dogs on "old old" human remains. In Kentucky, she noted, you can legally bury "your brother, your mother, your father" in the backyard. Training dogs to alert on old graves is "insignificant" and even distracting for the medical examiner's office when they are out on cases.

I understand her point. I remember the time that investigators spent pulling a cairn apart based on Solo and another dog independently alerting, only to get down to ground level and find roots that clearly had been there much longer than two years. If someone were farther down, it wasn't the victim we were looking for. They didn't dig. That was fine with me, though I remain mildly curious about that pile of stones overlooking a pond in the middle of the woods.

Increasingly, people are searching for historic human remains purposefully, using family Bible records, land deeds, oral history, Google Earth—and dogs. In the last decade especially, dogs have been used to discover or pinpoint what are essentially open-air museums: old cemeteries, battle sites, archaeological digs. One of the first documented uses of dogs on ancient remains belongs to the now-deceased bloodhound trainer and handler Bill Tolhurst, who in 1987 took his chocolate Lab, Candy, to an archaeology site in Ontario, Canada, after construction workers found a skull. Archaeologists realized the remains were from the War of 1812. Bill and Candy helped them locate three additional bodies.

Across the United States—from suspected massacre sites along the Oregon Trail, to hasty burials along the Old Spanish Trail, to slave graves, to Revolutionary War and Civil War burying grounds, to the prehistoric mounds of the Mississippi Delta Indians—archaeologists, historians, and geologists are teaming up with cadaver-dog teams to map where the dead might lie. I say "might" with deliberation. Only excavation and good testing can establish what lies beneath. Often excavation isn't possible. Or desirable.

. . .

"Thomasville, once simply the end of the railroad line in Georgia, has always been a well-kept secret because of its remote location," a *Road & Travel* magazine article observed. But to the fabulously wealthy industrialists who flocked there after the Civil War—the Vanderbilts, the Goodyears, and the Hannas—it was no secret at all. They all bought plantations and mansions at fire-sale prices. Cotton plantations became game-bird-shooting plantations after Reconstruction. By 1887, *Harper's* magazine had named Thomasville one of the top winter health resorts on three continents, with its salubrious dry air and increasing wealth.

"Northern beef and good fresh milk can be had here," the *Harper's* feature noted. After the encomiums, the writer offered a caution: "The popularity of this place makes it important for visitors to see that its sanitary arrangements keep pace with its growth."

Indeed. Two decades before, Thomasville's sanitary arrangements hadn't kept pace with its sudden growth. In the last throes of the Civil War, a panicked Confederacy, anticipating General Sherman's advance through North Georgia, shipped five thousand Union prisoners from the notorious Andersonville prison camp to Thomasville. Slaves in the small town hastily dug long trenches, six to eight feet deep and ten to twelve feet wide, to define a five-acre spot in the piney woods. The phalanx of Andersonville prisoners, ill, starving, and near death, were put in that hastily built camp prison.

The prisoners lived—and reportedly five hundred died—in Thomasville during twelve days in December 1864. The deaths were mostly from smallpox and diarrhea, and the numbers might have been higher if it hadn't been for the relative kindness of the locals. Physicians who already lived there and tended to the wealthy set up a temporary hospital in the nearby Methodist church.

Then the nervous Confederacy, realizing that Sherman had taken Savannah just two hundred miles to the east, moved the prisoners out of Thomasville. Those who survived arrived back in Andersonville on Christmas Eve.

The prison camp barely registers as a blip in the history of Thomasville or the Civil War. Though the federal government made the most sustained effort in the history of the country to disinter, identify, and reinter Union soldiers in federal cemeteries, it missed Thomasville. As Civil War historian J. David Hacker noted, "Men went missing; battle, hospital and prison reports were incomplete and inaccurate; dead men were buried unidentified; and family members were forced to infer the fate of a loved one from his failure to return home after the war."

The lot on Wolfe Street is tiny. Less than an acre of the original five-acre prison is still undeveloped, a patch of scrubby barren grass and a few pines and deciduous trees, surrounded by houses and city buildings. Two sides of the four-sided ditch survive, now L-shaped and sloping. A small historical marker notes the spot's significance, but the marker is dark and the lot shaded. I could find only one obscure guidebook that included its presence. That's in comparison with the hundreds of mentions of Thomasville's glories: its magnificent vacation homes and its huge oak tree.

Lessel Long, a Union soldier from Indiana imprisoned there, wrote at length about the Thomasville citizens, who he believed "manifested much sympathy for us." He also wrote of the terror sowed by the bloodhounds of the South at Andersonville Prison. These were, he said, the dogs brought from Cuba used to track slaves and to track Union prisoners trying to escape. "There is no doubt but what thousands of our men would have made their escape if it had not been for the dogs . . . They deterred many from making the effort to escape."

A century and a half later, a different kind of southern scent-detection dog would play a more benevolent role. The idea of bringing in cadaver dogs started when assistant city manager Kha McDonald, born and raised in Thomasville, realized she wanted to know more about that scrubby site with the small plaque. Thomasville had avoided the worst damage from the Civil War, but mysteries remained. Hundreds of Union dead were unaccounted for. Where were the dead who were treated at the Methodist Church buried? Was there a mass grave near the Wolfe Street site? The legend was that Union soldiers were buried under Broad Street, some distance away. That uncertain history was part of Kha's own legacy in a town built on slavery. At one time, the slave population outnumbered the white population in Thomasville. The city has a park named after the first black graduate of West Point, Henry O. Flipper, the son of slaves. "You can't escape that," Kha said.

A historian by nature, Kha decided this part of Thomasville's Civil War history, as well as the historic black cemetery, needed sunlight cast on it. She learned about cadaver dogs through a town librarian who was with a Florida search team just south of Thomasville. Dogs, Kha learned—including Suzi Goodhope's dogs—were being used in the Mississippi Delta to find 800- to 1,200-year-old human remains from the mound-building civilizations that lived there. Kha got in touch with Suzi, who connected her with cadaver-dog trainer Lisa Higgins.

Historic human remains weren't an obvious choice for Lisa. She had plenty of criminal and missing persons cases to deal with and a grueling seminar schedule. Lisa also admitted that in the beginning, she was deeply skeptical that the dogs were capable of detecting ancient remains. She has been converted, partly by watching her own dogs alert on remains going back more than 800 years, partly from watching other top handlers work their dogs. In several instances, she's received clear confirmation from excavations.

So Lisa brought Dixie and Maggie to Thomasville. Suzi brought her two Belgian Malinois, Temple and Shiraz, or, as she calls them, "the guttersnipe and the princess." Temple was a shelter rescue with post-traumatic kennel disorder and probably only part Malinois. Nonetheless, she carried many of those genes: high drive, opinionated, and hardheaded. Shiraz is like a piece of delicate, expensive china, also opinionated. With a fine nose. Shiraz's father won best of breed at Westminster.

Kha contributed her amateur historian's passion, as well as a geologist with ground-penetrating radar. Ephraim Rotter, curator of the Thomas County Historical Society, provided documents.

For Suzi, the work at Thomasville was fascinating and good for her dogs. It's not that the stakes are lower in these cases than for more recent missing persons, but they are different. Getting it right matters. Suzi noted that it takes time to imprint the dogs on "the older stuff." "To me, they work a little harder and a little slower."

Looking for the long gone is not straightforward. As Lisa Higgins noted, "Scent is all over." We humans might think scent would be strongest down inside the coffin-sized rectangular depressions that seem to indicate where bodies were buried in old cemeteries. That isn't always the case. As Lisa said, we can't know exactly where the dogs are smelling scent the most strongly. Low spots gather more scent. Animal burrows can make the scent more accessible in one area rather than another. Where roots engage, the scent can travel, although the means by which it does that are unknown—and controversial in both the handler and scientific worlds. We don't know all the mechanisms that make vegetation and roots more attractive to cadaver scent, although moisture may play a role, roots breaking the soil surface may play a role, and even certain compounds may become more available through vegetation. At this point, dueling and contentious theories abound. What's clear to everyone is that dogs appear drawn to vegetation and trees near a burial.

At the Wolfe Street site, the dogs moved along the partly filled-in ditch, slowing and alerting within a few feet of one another, in the same general area. Suzi lost count of the alerts. Kha, watching the dogs work, saw the pattern. "Suzi and everybody had just significant hits in the trench," she recalled. Then the dogs would go over to the property line. There's one tree there, on the embankment with a hole in the bottom of it. Every dog indicated enthusiastically there, Kha said, acting as if they'd "hit the glory land," as if someone were "blowing the fumes up" from the earth below.

When the ground-penetrating radar operators came in, they confirmed anomalies and soil changes in the ditch where Suzi had flagged dog alerts, as well as on the scrubby flat grass.

None of this is a scientific certainty. Kha knows that. But it's enough, combined with the historical records, for her to try to get a grant, even in this tight economy, to bring the obscure site out of the shade. Perhaps the town will add a fence or some markers. Many prisoners died on that site. If the massive oak tree in downtown Thomasville gets recognition for being the bigger oak in the Southeast, perhaps the final days of the Union prisoners in Thomasville will finally get some acknowledgment.

* * *

All of us, five middle-aged white women, looked at the open land in front of us. We were in West Virginia, 780 miles to the north of Thomasville, Georgia. A hill's curves ended at a pond half covered with duckweed. A red-winged blackbird trilled "conk-la-ree, conk-la-ree" before dropping off the power line to the reeds below. The hill was awash with blooming Bermuda grass, white and pink clover, a sprinkle of horse nettle. Streaks of green shot across the crest of the hill and just below the crest. A bull thistle interceded here and there. The six slaves reportedly were buried without coffins and not too deep. Maybe a child or two was there. A few might have slabs of granite on top of them, thought the farmer who owned the field, to keep animals from disinterring them.

It had taken a half hour for Kathy Holbert and her good friend and fellow cadaver-dog handler Lisa Lepsch to negotiate the search with the old man who owned the cattle farm, with its pristine white house and barns. Some of that time was taken up with listening to his memories of cattle drives from Montrose to the slaughterhouse in Elkins when he was young. About the two herding dogs who didn't mind getting kicked in the face and would challenge the recalcitrant back into line. His master herder, a long-haired red dog, was scarred and tough as nails. Probably not an Irish setter. Much of the talk was about the rules of dog engagement: that the gates be opened and closed quickly. That the dogs not stress the herd. Whatever the outcome, he said, he didn't want any archaeologists digging. If the bodies were there, they should be left in peace.

This site, unlike Thomasville, had no surviving documentation that we were aware of—if, indeed, anyone who could have created a record had cared enough. The oral history of the slaves, three men and three women, who perhaps lay on this particular West Virginia hillside in unmarked graves, had already served a modern purpose as a small branch of a bigger protest against a highway. The stories of the buried slaves—combined with endangered flora and fauna, Confederate and Union historic sites—helped prevent the huge ribbon of Route 33, a Senator Robert Byrd pork barrel project, from being laid across a section of Randolph County. The highway ended at Montrose and began again beyond.

When slaves were there, the top of the hill would have been filled with apple trees. Apples always played a big role in West Virginia agriculture. Any orchard would have had at least five or six varieties:

for cooking, for eating, for cider, for preserving, for shipping, dried, to England.

Digging a grave in that orchard would have been difficult, with tree roots and rocky soil interceding. That was what West Virginia had: rocks and apples, salt and coal. What it didn't have was a lot of slaves; it was a mountainous region with few large plantations, and tobacco grew only in a few areas. Cotton nowhere. Other than the salt mines, where slaves like Booker T. Washington toiled, most West Virginia slave owners were farmers with fewer than five slaves.

Just because slaves were relatively few in number didn't mean they were treated humanely. Some scholars believe that the treatment of Appalachian slaves was worse than on many Deep South plantations, with accounts of much harsher physical abuse and families torn apart more readily. Far too many scholars, one researcher wrote in frustration, "confront me with the mythological construct" that Appalachian slaveholders were "small farmers who only kept a couple of slaves to help their wives out in the kitchen."

The farm's owner, ninety-two, told Kathy and Lisa that the slaves hadn't been allowed to leave the farm. Some of the children, he said, had been sold. If that were the case, it was probable that they walked in a coffle, bound together with ropes, a frequent sight in West Virginia in those decades. Nearly one in three slave children in the upper South in 1820 was gone by 1860.

It was just after nine A.M., and the sun was glaring when Kathy started Strega loping down the cow path. Strega's rear moved sideways due to her aging hips as she went toward the pond, folding back like sable-colored origami. She moved along the outer edge of the pond, not bothering to dip herself and not slowing until she hit the end. She held her nose low to scoop scent. Nothing here. Nothing here. No reason to quarter or hesitate. Ten yards later, she slowed and flipped back on herself in that distinctive way a scent-detection dog does when it notes something relevant. She leaned over a drainage spot. I could see, tracing back up, where the water percolated off the rocky hill and down into a dimpled spot at the bottom before entering the pond. Strega paused there, sniffing, turning around several times, leaving, then going back to it. She never looked at Kathy, who was standing back about a hundred feet, silent, letting her work. Kathy is a noninterventionist handler. Strega left the drainage and slowly worked up the hill toward its crest, using her good, deep nose. This is behavior that experienced handlers and trainers have seen with dogs working toward victims, even those who are not buried. The scent can percolate down the hill or toward water from a great distance. Scent won't always be strongest right on top of a grave.

After more than 175 years, one shouldn't expect vegetation to be different above the buried, Kathy said. Nonetheless, at the top of the hill, a slightly concave area was filled with purple and white clover and mats of shorter, greener grasses that seemed woven more tightly than in other places on the hill. Strega stopped, turned in a tight circle several times, and whuffed audibly. Then she sat and stared at Kathy.

And so it went for the next two hours. Renzo, Lisa Lepsch's massive black shepherd, found a spot on the other side of the small hill. It wasn't concave, but it stretched out for fifteen feet or so in a green streak across the top of the hill. He sniffed around the edges of the space, walking slowly, then came into the middle of it and stood there looking at Lisa, who chuckled. "He's good at defining things," she said.

Rocco, Ann Christensen's young shepherd, alerted where Renzo had shown interest, flinging himself down in abandon. Whump. Dusty, Charm Gentry's Beauceron, alerted there, too, after mincing through the field and occasionally popping straight up like an antelope that had found a snake beneath her. So did Kessa, Ann's gray sable shepherd.

On that hill of four or five acres, the five dogs walked and ran, quartered back and forth, slowed, then alerted on two general flat spots where Strega and Renzo had alerted. Dogs can alert on top of other dogs' spots, although that's not how they are trained. Handlers can subconsciously signal to their dogs. None of the handlers was working the problem blind. While the dogs and the oral history overlapped, using GPR in that rocky environment would be impossible. Once we left the field, all that would remain would be the stories and GPS locations marked in the dogs' training records. No names. No headstones. A field of clover and thistle.

"I've never done anything that old," said Arpad Vass. Then he continued. "But it does not surprise me that a dog will alert. Clay makes a nice vault."

. . .

There's little science to shed light on what is happening when dogs sweep through an old graveyard. Archaeology is already a speculative discipline. What can dogs' noses add? What are they alerting on? Handlers' unconscious cues? Old decomposed trees? Gravestones, when they are there? Are people simply watching the dogs work, adding historical fantasy to evocative landscapes, and creating erroneous paint-by-noses pictures?

Several teams across the country now work with old burials; a few are starting to get consistent and more verifiable results with the combination of ground-penetrating radar, dogs, and oral and written history. Scattered excavations here and there—proof positive—are corroborating those finds. The arena of cadaver dogs and historic remains is still clouded and contentious, though. It's particularly hard to prove the worth of working dogs when dealing with old burials, because more often than not, excavation (aka confirming proof) just isn't going to happen.

Mary Cablk has done research using historical-remains detection dogs on the Old Spanish Trail on several possible burial sites, including some blank areas. Now she would like to corroborate, with core sample testing, where the dogs alerted and where the GPR showed changes.

"The historical human-remains detection is something that I go back and forth on," she said. "Do I think that dogs can do it? Yes. Do

I think that all the teams out there across the country who claim they can do it can do it? No. That's what I call faith-based dog work." Are dogs trained to detect ancient remains, as Mary notes when she is at her most skeptical, simply "great anomaly detectors"? She wonders whether the anomaly might be simply a change in the surrounding soil chemistry rather than ancient remains. Are the dogs alerting on the pottery urns that human ash was placed in, in some cultures, rather than the ash itself?

Mary is not the only skeptic and not the only researcher trying to develop peer-reviewed research on the issue. The work on using dogs for archaeological work is just beginning, and it's all over the map, literally—from Bosnia to Hawaii, from California to the Mississippi Delta. Since no one has established exactly what volatile organic compounds cause dogs to alert for more recent deaths, old graves raise even more questions.

"Dust thou art and to dust thou shalt return" has a scientific basis. At some point, most of us rejoin the earth in a way that should confound a cadaver dog as well as any instrumentation: We're not grave dirt, we're not dirt mixed with adipocere. We're just dirt, plain and simple.

How far back can the dog go? It depends on how much scientific verification we want. Solo has alerted on an eight-hundred-year-old bone from the Mississippi Delta, and I've watched a number of dogs do the same. And, of course, it depends on what the dog is actually alerting on—it was thought until recently that it could only be VOCs. The human body simply lets volatile organic chemicals go up and up, until the body stops communicating in that fashion. No scent understood as human decomposition should still be holding forth into the air column. And yet, good dogs, trained on the whole spectrum of decomposition, appear to know. The soil seems almost permanently changed.

"Scientifically, it's almost impossible to explain," Arpad Vass said. "It's long gone, so what could they possibly be picking up?"

Scarcely had she spoken, when a stiffness seized all her limbs; her bosom began to be enclosed in a tender bark; her hair became leaves; her arms became branches; her foot stuck fast in the ground, as a root; her face became a tree-top, retaining nothing of its former self but its beauty. Apollo stood amazed.

—Apollo and Daphne, *Bulfinch's Mythology*, 1913

When I watch well-trained cadaver dogs work a possible clandestine burial site or define the outer perimeters of old cemeteries—throwing their heads, staring up into the trees, even putting their feet up to try to climb them, and bringing their noses deep down into tree roots growing out of depressions in the earth—I'm fascinated. I tread more lightly in those spots. As one handler in Mississippi, Gwen Hancock, whispers when she accidentally steps in the shallow depressions of the nineteenth-century cemetery she and other handlers discovered by accident in the woods behind her house: "Please forgive me. Thank you for letting us train our dogs."

I wanted to say similar thanks when I got invited on a small expedition to a possible cemetery site in South Carolina. I had watched Kathy Holbert and Lisa Higgins and a number of other experienced handlers work their dogs on possible grave sites in West Virginia and Mississippi; Solo hadn't been with me. Then cadaver-dog handler and anthropology graduate student Paul Martin invited us to South Carolina.

Early-morning mist was still hanging just above the Great Pee Dee River on the border between North Carolina and South Carolina. A February mist with a bite to it. On a little bluff overlooking the river stood one lichen-covered stone, placed there by the Daughters of the American Revolution some time in the 1960s. The DAR had hoped, in planting the stone, to mark an abandoned cemetery that might include the body of Revolutionary War Captain Claudius Pegues Jr., who fought alongside the wily Francis Marion. Marion, the Swamp Fox, led a group of backwoods soldiers against the British, often escaping into the underbrush or marshes—which was how he got his nickname: "As for this damned old fox, the Devil himself could not catch him," swore British Lieutenant Colonel Banastre Tarleton.

Claudius Jr. died in 1792, less than a decade after the war of independence was won. He was reportedly buried here. The bluff overlooks the once-huge Pegues family cotton plantation, the site of the only prisoner exchange during the Revolutionary War. But perhaps Claudius wasn't here at all. Putting a stone marker on a spot doesn't always make it so.

Claudius Jr.'s great-great-great-great-granddaughter, Pat Franklin, stood at the edge of the woods, her gray hair swept up in a loose Gibson-girl knot. Pat wasn't familiar with cadaver dogs; nor was her longtime friend and fellow genealogist May MacCallum. But after May read about dogs finding graves, she did her research, and ultimately, they called Paul.

Pat was uncertain how the family records, letters, wills, and oral history coalesced. Her grandmother always told her "the old burying ground" was at the Charrows, where we were now, looking over the river, and that a coach whip snake guarded the spot. A coach whip snake is long, fast, and smart, with large dark eyes and scales that look like braided leather. It will chase you, whip you to death with its tail, and then stick the tip of its tail up your nose to make sure you're not breathing.

Childhood fears gone, Pat wanted to know where her ancestors lay. Claudius Jr., his wife, Marcia Murphy, and perhaps four infant children might rest in front of us in the woods. Two children died the same year they were born; two were dead by the time they were two. Had they lived, they would have been heirs to a large cotton plantation. Their wealth didn't spare them the early death that greeted so many infants. Some took their mothers with them. One record indicates that Claudius Jr.'s mother, Henriette Pegues, may have been buried here just days after she gave birth to an infant girl in 1758. The records don't show what happened to the little girl, but Claudius Sr.'s will mentions only two sons—directly, that is. Claudius Sr. gave his servant woman Cortney and her son, Martin, to his son, William, as slaves. He also gave Cortney her own slave and directed that a house be built for her and Martin. He stipulated that the two of them were to be set free when Martin was twenty-one, and that Martin be given tools and taught a trade. Cortney and Martin then disappear from the record.

I couldn't quite countenance, looking across the land, that both the Revolutionary War and the Civil War took place here, that people who had fought in the revolution for independence from England were the same people who kept slaves. And that their slaves kept slaves.

Solo whined loudly in the car. He was a cemetery novice and would not get to go first, which irritated him greatly. Paul Martin would start with Macy, one of his veteran cadaver dogs. Paul started working on ancient human remains more than ten years ago, and it has become both an intellectual and a training challenge for him. A bit more sun, Paul said, and the scent would start rising and moving, if scent were there. Too much sun can burn it off. There's a sweet spot for grave work.

Arpad Vass observed the following about using dogs to find more recent, clandestine burials, and Paul was applying the principles to older remains: Humidity should be between seventy and eighty-five percent. Check. Ideal soil type: sand and humic. Check. Temperatures at fifty-three degrees and rising. It was colder than that, but otherwise, we were searching in conditions that were well-nigh perfect, according to Arpad's calculations, although I didn't know if the barometric pressure was falling, as it should be.

Macy looks more like a cross between Old Yeller and Gollum than a Labrador. He's slippery and primitive, with amber eyes, a reddish-dun coat, and ribs sticking out; though he eats constantly, he runs it off. He had raw spots on his pink nose that morning from butting his wire crate door repeatedly in his eagerness to get out. Macy banged in joy as Paul approached the crate; as soon as the latch turned, Macy shot out into the woods. Paul followed more slowly, then stood amid the oak, beech, sweet gum, black cherry, and sycamore trees. Calm and quiet, he watched Macy dash around the perimeter of the area like a surveyor on methamphetamines.

"Too far," he said in his nasal, lilting voice as Macy dropped out of sight down the bluff toward the river. It's a term that Lisa Higgins uses, and I had started using it with Solo. It doesn't mean "come." It means "start to circle back."

Macy was working hard, snorkeling scent in the leaves without finding anything. It was too cold. He ignored Paul and didn't ask to be rewarded. After ten minutes, Paul put Macy up. We would have to wait until the sun penetrated the canopy. In the meantime, another glade beckoned, where it was warmer than it was on this crest. Down the gentle hill, where the cotton was harvested months before, lay the "black cemetery"—covered with periwinkle and abandoned but with stones that dated through 1910. A number of former slaves became tenement farmers and stayed during Reconstruction and beyond. Pat and May have worked to record this cemetery as well.

Paul turned to me. "Why don't you go get your dog?"

Solo was ecstatic to be out of the car, whirling and barking sharply, coming back to hit my leg. I offered him water, since he had been panting in excitement. He wasn't interested. He dashed down the hill, ignoring the rough cotton stubble, and into the woods and underbrush and tangled periwinkle at the bottom. Scent must have called to him. By the time we entered the woods, he was already working. Solo had never been exposed to gravestones, except for a hasty search of a modern graveyard; he had no reason to suspect they had any significance. Given how they were scattered, neither did I. Nonetheless, in one spot, I could hear Solo snorkeling from twenty feet away. He did his down alert, staring at me. He was surrounded by periwinkle and daffodils, and I saw several tilted headstones under the vines. "Reward him," Paul said. I did. Solo alerted seven or eight times. Paul estimated that at least seventy-five people were buried at this larger site. Pat doesn't know how far it goes back. The DAR has not put a stone marker here.

It was now midmorning, so we moved back to the top of the hill, where we expected fainter scent because of the age of the burials, if they were there. Paul took out Macy again; ten degrees warmer, and Macy's behavior change was astonishing. He alerted several times. We started to notice shapes and depressions that had been covered with leaves. Paul pointed them out to Pat and May. The more oblong and rectangular, the more likely. Possibly up to six adults and five children, Paul estimated.

Paul then ran Jordan, his other Labrador. She was soft and black rather than hard and amber, like Macy. She alerted repeatedly in the deep leaves, in a couple of spots where Macy had alerted, in a couple of new places, where we could then see depressions. Both dogs worked the area in ways so similar that I marveled. Macy was fast, Jordan much slower. Nonetheless, the same depressions seemed to hold scent. The two Labradors threw their heads around the same trees, went down to the same holes, and did their final alerts in four or five places.

I could hear Solo howling. Paul turned to me, and I freed Solo from his Camry prison, letting him run into the little woods. I stood well back. It didn't matter. His work overlapped Macy's and Jordan's—several alerts and head throws in the same places. I was no longer surprised.

We had finished searching for the day. We placed flags, and the work of measuring began. Pat and I went to poke the depressions shallowly to see if we hit stone; markers could have become buried underneath the humus of the woods. Burials of the era tended to be at least four feet deep, so it didn't feel as though we were poking at the dead.

May was happy for her friend. "We don't care who is who. Now we can lay it to rest." Pat and her family were planning to put markers there.

Solo was lying off to the side, panting, his tug toy in his mouth. The leaves where he had flung himself were disturbed. Beneath them, I saw a hint of white. A tiny violet. Pat looked at it and smiled.

"It's called 'spring beauty.' "

17

A Second Wind

My little witch, who lived life every day, has traveled on, taking a huge part of my heart with her. Fair winds and following sails, Strega. Buon viaggio.

—Kathy Holbert, 2012

Sweat poured down Danny Gooch's face. He'd just removed the suffocating decoy suit and rolled up his soaked dark blue T-shirt sleeves over the tops of his shoulders, exposing the dark-blue-inked portrait of a Dutch shepherd head on his bicep.

One of the handlers shook his head in dismay. "Paisano," he said.

Danny's flat white teeth flashed as he pulled back his lips in a grimace. "Hey, you know what Kimbo would have done? You know what Kimbo would have done?"

Everyone standing there knew what Kimbo would have done. Earlier, Mike Baker had decoyed, stashing himself underneath the composite tile floors in an abandoned laboratory building. Watching him thread himself into the small space, more cramped than a coffin, gave me claustrophobia.

Dogs are object-oriented. Once the less-experienced patrol dogs realized Mike's scent was somewhere in the room, they went to look for him behind doors. One leaped repeatedly at the large refrigerator unit in the room; Mike's scent had crawled up its side. Several other dogs figured an interior hallway window was the key. They kept levitating toward it, hoping that a person would materialize on the other side. The idea that someone could hide literally under their paws flummoxed them.

Dark little Kimbo, who looked more like a Tasmanian devil than a Dutch shepherd, would not have been fooled. He would have smelled Mike in the small holes in the particleboard, then tried to dig him out of the floor with his claws and teeth, with Danny making sure he didn't entirely succeed. Mike muttered one night, while watching Kimbo enter a deserted cigarette factory in downtown Durham, stiff-legged, ready to rumble: "If you're going into hell, that's the dog you want in front of you."

Kimbo wasn't there that night to show them how it was done. Danny retired him when he was twelve years old, older than most patrol dogs. Kimbo's toughness kept him going; he was finding dropped guns until the end.

Sometimes retirement is planned, as it was in Kimbo's case. Danny's new patrol dog, Rin, a handsome dark German shepherd, is a third again Kimbo's size, a happy and all-around fine dog. Danny, a Malinois and Dutch shepherd snob, had to dial back his faux anti–German shepherd rhetoric—too soft, too slow, too big—although because it was Danny, the dial didn't move too much. Kimbo stayed at home in an increasingly customized kennel space, getting spoiled by Danny's daughter, who sneaked him into the house when Danny was at work.

Waiting too long for a dog to retire is a big mistake. It shows when K9 officers look with almost tender pity as a dog's back leg starts to shake when it should be solid, or when a decoy gives a dog less of a swing on the bite sleeve because he's worried about hurting the dog. That protective feeling can get officers injured.

Sometimes retirement comes tragically fast. One hard-charging five-year-old Malinois in Durham came up lame, and within two weeks he was diagnosed with degenerative myelopathy, a disease that destroys the spine and shortens the life of working dogs. He was off the force a few weeks after that. He would never again vehemently thrust his Kong toward K9 officers who reflexively and protectively cupped their hands over their crotches when they saw him arrive.

Down in Fort Lauderdale, Florida, Steve Sprouse retired his blocky-headed German shepherd, DJ, at eight because of arthritis and spine problems. When DJ lunged forward to hit the bite sleeve, it was clear he was in pain; he yelped when the jolt traveled down his back. He needed help getting into the back of the patrol car. Steve retired him just before I traveled to Florida to watch Steve training.

"He gets out of the patrol car just fine," Steve's wife, Sandy, pointed out, looking at the shepherd fondly. DJ was tearing up coconuts from the backyard palms with Casey, their wild young sable female, as company. Sandy has lived with every one of Steve's patrol dogs over the decades, through their retirement.

DJ's forced retirement had Steve wondering as he tried on a stiff new bulletproof vest the department had issued him: Was he too old to start another dog? Were his reflexes on the street still good enough? All the moves that come naturally on a deployment would need to become part of his body memory again. He would have to synchronize with a new dog. Sandy and he would have to squeeze a third large, energetic dog into their modestly sized house. "Do I have another dog left in me?" he asked.

It's a familiar question for handlers of a certain age, and not a rote one. For most professional K9 handlers, losing or retiring one dog becomes decision-making time. Some handlers get into K9 not because they love working with dogs but because of the prestige. They realize only afterward how all-consuming life with a patrol dog is. Sometimes promotion or another division can beckon. But I know more than one person who refused promotions in order to continue working with dogs in a unit.

Sometimes a handler's own retirement beckons. Steve, who spends most of his time training others, had some thinking to do. A new dog was more complicated than a new vest.

. . .

Solo's left hock can wobble, particularly the day after a demanding training or a long search. His right front shoulder appears arthritic, probably from years of leaping off the fourth step each time he comes downstairs. He slips occasionally on our concrete floor as he waltzes back and forth across the house, shaking his toys until they are dead, a nightly ritual before heading to bed.

Except for a smattering of gray on his muzzle and lower jaw, Solo doesn't look his age. In any case, German shepherd muzzles can get gray before they turn three. He doesn't act his age, either. He still dolphins around the yard, brings us toys at night, head up, a gleam in his eye, setting them on the couch, pushing them with his nose, and then backing off and crouching down slowly to see if we'll take the bait he's carefully set. When company arrives, he becomes a tiresome clown. When I tell him to go lie down, he'll run at the dog bed and jump on it with his front feet so he can use it like a boogie board to skate across the floor before flipping in crazy circles and throwing himself down on it with melodramatic yowls and moans. You're killing me here. Guests laugh, escalating the noisy theatrics. If a particular find during training poses a challenge, he'll self-reward by doing a couple of extra victory laps, snaking through trees and leaping over obstacles, swinging the tug toy.

On searches and during training, he is still capable of being a jack-ass. During a recent search, two large dogs chained up in a backyard were too much of a temptation. Solo minded me but at the last minute; that's not sincere obedience.

It's rare these days, though, that I get truly angry at him or even embarrassed. It's not that I've gotten softer; it's that he has gotten better. And it's a sign of the beginning of the end. A summer training at Nancy's ended with her throwing up her arms as he panted into an indifferent alert on a cadaver hide, tongue hanging on the grass. "Just pathetic," she declared. She looked at him with scorn and then at me accusingly. "You really need to start another dog," she said, "if he's going to be around to teach it anything at all."

I searched for excuses. It was, I said with triumph, the prophylactic antibiotics he'd been on for a bad tick exposure. Maybe it was his boredom with her hides. Same time, same place. Same dead stuff. Been there. Smelled that. Nancy knew my protestations were a delaying tactic.

Working dogs depend on their fitness and strength. At some point, climbing the hill at the solid-waste landfill to follow an evocative scent coming from the other side will hurt too much. The best nose in the world attached to arthritic hips or legs won't cut it.

I spent the first three years of Solo's life wondering which full-bore accident or macho posturing might kill him: running at top velocity into a barbed-wire fence on a search and bouncing off it like a bird hitting a window; deciding to take on a herd of cows; charging out of a warehouse and belly-flopping off the loading dock onto the pavement below. He survived those and many more incidents of our combined bad judgment. Then we reached homeostasis for a few years. He was at the top of his game mentally and physically, no longer as impetuous.

When a big water moccasin came thrashing out of the murky water and toward us during one search, I simply called Solo off and we moved on, leaving the snake to its territory. We could search pastures filled with horses and I didn't worry. I could speak in low, quiet sentences and mostly not at all—unless we flushed a coyote from her den or a fawn out of a swamp.

Those years were lovely and temporary. Solo is now eight, the age where physically, things start to go wrong no matter what. Like me, he is headed downhill. And because he's a dog, he's moving faster than I am. As always. Dogs don't last. At best, he's got just a couple of years of work left. How can I know? By being ruthlessly honest and not telling myself stories about old dogs and their wisdom. Not just joints and muscles suffer from age. A dog's sense of smell is not forever. The nose's abilities diminish with age, with disease, with a series of micro-injuries, Larry Myers noted. People's noses aren't that different. Human scent expert Avery Gilbert wrote that human olfactory performance starts to deteriorate when people are in their forties. This nose degradation story comes with caveats: Perfumers, Gilbert said, get better with age. "A given seventy-five-year-old may outperform a given twenty-five-year-old. . . . Experience and skill more than compensate for any dimming of acuity that comes with age."

The same can be true of dogs, according to Deak Helton. Experience, skill, and good physical condition can help a dog compensate for certain kinds of aging. I've seen it with several of the Durham Police Department's patrol dogs, since I've watched them now for more than six years. A couple of them, around Solo's age, simply get the job done. Perhaps not as dramatically, at a frenetic run, as they did a few years before, but with such admirable efficiency and clear knowledge that it's not until they're back in the patrol cars that you realize the older dogs finished the job in half the time.

Watching those dogs get older made the issue of "retiring" Solo seem overwrought. What does it mean to retire from part-time volunteer work? All I knew was that I wasn't ready to quit.

· · ·

The small lump on Solo's leg, just above his dewclaw, appeared a few months after I noticed his occasional limp. I called the vet immediately, but by the time David and I got there, I had talked myself into the obvious diagnosis—bone cancer. Osteosarcoma in German shepherds, more frequently seen in males than females, is common. It shows up on those long leg bones at a median age of seven and a half. By the time a limp appears, the cancer has usually metastasized to the lungs. Cure rates are low.

I assured myself and David, trying to ignore the gaping nausea in my chest, that Solo had had a good life, an active one, and while it would be truly sad to lose him, it wasn't a tragedy. We'd had a nice run. I laid out the limited treatment options in my head, to be ready for the vet's arguments. We'd do surgery, sure, but no amputation, and no radiation or chemo.

David and I had already had this discussion about Solo's shortened life span because of where we train and search. That nose, those lungs, those feet and legs have been exposed to all sorts of crap. Swamps and fields filled with runoff, herbicides, pesticides, heavy metals. Abandoned houses filled with chemicals and lead paint. Wrecking yards with heavy oils and antifreeze slowly oozing into the ground. One time we worked on fields that had just been sprayed with sludge— human waste treated with heavy doses of chlorine. After that, I had a wide-ranging conversation with a friend who is a public health epidemiologist. He knew a lot about sludge. Sterilized human poop is somehow more disgusting than its original form. To say nothing of human garbage that has steeped in landfills for decades until it oozes out as corrosive and toxic hydrogen sulfide and sulfur dioxide. More than a few times, I came home with heart palpitations from having trained in dusty tobacco warehouses and factories, wondering how I'd managed smoking on and off for nearly twenty years before finally quitting, and whether Solo would get lung cancer.

My cure-all for toxic exposure is soap and water: first Solo, then me, as soon as we get home. Solo, who has just spent his time throwing himself with delight into muck-and-algae-filled swamps, will tuck his tail and lay his ears back tight against his head, trying to seal them against a single drop of clean water.

Is Solo more prone to cancer or respiratory illnesses or stray bacteria than a standard pet dog? It's hard to know. Few studies are out there. Working dogs tend to be in good physical condition and less obese than pet dogs. It may go beyond that, though. Cynthia Otto, founder of the Penn Vet Working Dog Center, studied the dogs deployed during 9/11 and came away encouraged. She told Terry Gross of *Fresh Air*: "[T]hese dogs seem to live longer and healthier lives than your average dog that we see in our hospital at the University of Pennsylvania. So I have a theory that I would love to explore as well. If the physical activity and the fitness and the mental stimulation and just the joy of life that these dogs have, because they're doing something so great and have such a great bond with their handlers, if that doesn't enhance not only the quality of life but their longevity."

Otto is not the only researcher to look at the dogs of 9/11, although she concentrated on the volunteer search-and-rescue dogs.

Another small study of the twenty-seven New York City police and fire department dogs who worked at the 9/11 sites showed similar results. In the five years after those dogs were deployed, their health problems were "minor and infrequent." The dogs were exposed to air laden with the dust of cement, glass, fiberglass, asbestos, lead, jet fuel, dioxins. Those dogs worked even longer hours than the search-and-rescue dogs who arrived a bit later to the site: thirty-seven weeks without masks, without hazard suits. They were getting toxins through every pore. None of the dogs showed long-term respiratory problems—the diseases that affected the human rescue workers at a higher rate.

The average dog in our industrialized nation isn't doing so well. One 2008 study of pet dogs showed they were contaminated with thirty-five chemicals, including eleven carcinogens and twenty-four neurotoxins. Happily, one can't leap from chemical contamination to disease, but skin cancer, bone cancer, and leukemia are much more common in dogs than in people. It is estimated that almost 50 percent of all dogs over the age of ten will develop cancer and approximately 20 to 25 percent will die from it. In the late 1930s, a researcher showed that chemical compounds used in dyes caused bladder cancer in dogs. In the mid-1950s, other researchers showed that another industrial chemical caused bladder cancer in dogs. By 1980, a large study of pet dogs showed that bladder cancer correlated with living in industrial areas.

I arrived at the vet's knowing all of this, resigned. After fourteen years of living with me, David was resigned to my fatalism, realizing that he couldn't stop my predilection for imagining a bleak future. Solo doesn't do resignation, so he struggled against the biopsy needle, believing the vet was going to do the unthinkable and trim his nails. Once he felt the thick needle piercing his skin, which was no big deal to him, he relaxed completely and let her take the cells and smear the slide. She didn't need a microscope for the first step.

The vet held the slide up to the light and then smiled. "Grease." The most beautiful word in the vet lexicon. Solo had a benign cyst.

· · ·

Solo, like most American volunteer search-and-rescue dogs, hasn't had constant exposure to toxins, compared with, for instance, the military working dogs deployed in the Middle East and South Asia. And the military working people deployed there. People come back with a variety of health problems that might or might not be related to what they were exposed to while serving.

Dogs are one of the best warning systems we humans have, and not only as guard dogs. Dogs can be both sentries and "sentinels"—an early-warning system for understanding human disease. The general idea of using animals as a human model in biomedical research goes back to the ancient Greeks. But the idea of sentinel animals, which because of their shorter life spans might help us understand how disease or toxicity affects humans, is a slightly newer medical model. The earliest recognized sentinels were the canaries in the nineteenth-century coal mines; they dropped dead from methane or carbon monoxide before the miners did. In 1952, cattle dropped dead in the smogs of England. Then the people died alongside them. In the 1950s, in the tiny fishing village of Minamata, Japan, cats started exhibiting bizarre behavior, "dancing cat fever." The cats sometimes fell into the sea and died, what villagers called "cat suicides." Then symptoms started showing in people. Every living thing in the village was suffering from mercury poisoning, coming from an industrial plant making vinyl chloride and sending its effluent into the bay.

Though dogs may be a great model for studying disease in humans, it is understandable that the military might not leap to investigate diseases that military working dogs suffer over their lifetime. In one of the few studies released on military dogs from the Vietnam War, both dogs and servicemen had elevated risks for testicular cancer. Hypotheses about the causes have ranged from exposure to Agent Orange or malathion to taking the antibiotic tetracycline.

Studies on the health of military working dogs are in their infancy. One epidemiologist privately expressed frustration about how little had been done. That is changing. A major study is under way on military working dogs deployed in Iraq and Afghanistan, and early results may be out by late 2014. Military personnel who served in Afghanistan and Iraq may have been exposed to environmental toxins. Sandstorms there are dramatic and last for days. The dust particles are inhaled. Burn pits that the military has used to get rid of everything from human waste to metals have created exposure to smoke that concerns veterans' groups. If service people were exposed, so were the military dogs working alongside them.

Michael Peterson, an epidemiologist and veterinarian in charge of the study for the Veterans Health Administration's post-deployment health group, said that the reason for studying the dogs is "99.9 percent human." Peterson and a VA colleague, physician Wendi Dick, whose specialty is preventative medicine, are focusing on respiratory diseases first and foremost, but also on cancer, neurologic diseases, and neoplasms.

Peterson, Dick, and their fellow researchers are looking at medical records for a group of 450 military dogs deployed overseas from 2004 to 2007: Malinois, German shepherds, Dutch shepherds, and Labradors. They will be compared with a control group, dogs who were destined to deploy but stayed behind in the Military Working Dog Center at Lackland Airforce Base.

If the military dogs died or were killed in Iraq or Afghanistan, the researchers have their tissue samples. Necropsies used to be routine for even healthy military working dogs, since they were standardly euthanized rather than being adopted out. That has changed since Robby's Law passed in 2000. Now military working dogs can return from overseas, be adopted, and get lost as a medical subject. It's hard, except in the abstract, to regret that change too much. Then there are dogs who were part of the contractor system and may have deployed for years at security points and embassies; there are dogs who, like Kathy Holbert's dog, searched for missing soldiers. Those health histories are inevitably lost or incomplete as well.

"We're really just testing a hypothesis. We don't know what we're going to find," Peterson said. "We're keeping the categories very, very broad. This may be a dead end. We may stumble across something. This is just another piece of the puzzle."

. . .

Dogs get sick, get old, die. Certain dogs you miss more than others. You try not to, but inevitably, that's what happens.

Kathy Holbert's cadaver dog, Strega, was eleven when she was diagnosed with bile duct cancer. She had been over in Iraq and Afghanistan for nearly a year. She became ill and died within a couple of months in the mountains of West Virginia, where she'd spent most of her life. Kathy did everything, including an experimental procedure that injects a dye to help the immune system respond. She knew it was a long shot.

Danny Gooch's Kimbo died in March 2012, a year or so after his retirement. Danny's daughter had the vet keep one of Kimbo's canine teeth. For Father's Day, she had it mounted on a small gold clasp. Danny wears it on a chain he tucks down his T-shirt. His daughter got a small heart tattoo: KIMBO, BFF. Kimbo was dark and fierce; his tooth is small and white.

Sean Kelly had four patrol dogs during his police K9 handling career. Nero was his most recent, one of the happiest working dogs I'd ever met. He loved people, finding drugs and guns, and biting. He had a deep, forceful bark and one of the smoothest on-and-off switches I'd witnessed. He'd come off the bite sleeve and immediately sidle up to anyone watching, stump tail wagging. Nero visited homes for developmentally disabled children where the only coherent word Sean could understand, joyfully shouted over and over, was "Nero! Nero!"

Nero's ashes sit on Sean Kelly's mantel in North Carolina, in an eight-sided canister marked with his paw print. The Malinois's print is also tattooed on Sean's big calf. Nero's ashes aren't the only ones sitting on the Kelly family's mantel; Sean has the ashes of two of his other patrol dogs there as well.

Losing Nero hit Sean the hardest. Nero had been a military working dog with the special forces. He'd survived an IED that had killed his handler. The same bomb blew out Nero's incisors and put a gaping hole in his chest. His tail was amputated because his hindquarters were so damaged. He was a mess, but he was alive. Veterinarians patched his body together and gave him titanium teeth. But Nero, who had been one of the top apprehension dogs in his unit, who had helped save his handler before, as well as other unit members, could no longer do his work. Post-traumatic stress disorder is too clinical a term: Nero and his world had been blown apart.

"You can't get into the dog's head and say, 'It's okay, it's not going to happen again,'" Sean said. Besides, that wouldn't have been the truth. So Nero returned to a military base in the U.S. to help train green K9 handlers. However, he had a lot more working parts than nonworking parts. Turning him into the equivalent of a practice dog was wasting a great dog. So in 2008, in a quietly arranged deal, Nero went to a police department in North Carolina. Sean, a former K9 officer from New York City, fell in love with him. Nero was six years old. He had drive, smarts, and a great nose. Although his tail and teeth were mangled, his head and chest mostly white, he still had his perfect, pointy Malinois ears. Inevitably, Nero's work in a small city in North Carolina, although not as dangerous as being with special forces in Iraq, had its challenges. Nero ran into a yellow jackets' nest tracking a suspect.

"I don't want that ear to flop over," Sean told the vet when he brought Nero in. Nero's head was swollen with stings. "That's the only thing he's got going for him." His pleas were in vain.

Despite his now comically flopped ear, Nero kept working. He got a second wind. He worked with Sean for three more years on the street. And while Nero had loved work, retirement at the age of nine felt fine, too.

"He adjusted too well to home life," Sean said. "I'd come home, and he'd be on the bed." Grinning, titanium canines gleaming.

The lymphoma was fast and aggressive. Despite chemotherapy, it was over in a matter of weeks. Nero was one day shy of his tenth birthday when he started gasping for air. Sean rushed him to the vet and into an exam room.

"I was sitting on the floor, and he came over and looked at me. I knew."

• • •

Solo and I faced man-made mountains in downtown Durham off a gravel road where the city keeps street repair supplies. Mike turned his SUV around and put on its high beams, and I did the same with my Camry's headlights, so they cast pale light on the hills of yellow sand, crusher run, and rubble at the end of the road. The piles created cantilevered half-pipes of sand and granite rather than snow. The valleys between were cratered by tire ruts filled with muddy, alkaline water.

The limestone crusher run looked one-dimensional, like piles of gray-and-black fragmented leaves or frozen dirty ice that went straight up fifteen or twenty feet, an Escher without open space or grace. The sand across from it was piled even higher, a huge dune stretching out into the dark. It had been unused for so long that animals had burrowed into it, making cave villages on the sides. Across from that lay a heap of granite curbstones; the city of Durham had cut and pulled the stones like long narrow teeth from the mouth of downtown. That granite had lined the streets for decades, five inches wide and three feet down, curbs that shredded the tires of those who are bad at parallel parking, and testified to a city of an estimable age. Immovable. Strong. Here, piled and canted in every direction, the pinkish granite looked unstable, with black gaps: seesaws or dominoes if Solo stepped wrong. The skunky smell of asphalt permeated the air here, though we were surrounded by pine trees and fragrant wax myrtle.

Mike had planted some training material for Solo, somewhere on the mountains of crusher run, or down the half-pipe of sand, or in the rubble pile.

Solo barked sharply at me, impatient to be released, and then disappeared briefly into the woods and darkness to give the trees his canine greeting. He came out of the dark, back around the edge, running smoothly. I didn't need to tell him to start work. He motored up the sand dune, ran across the top, disappeared along the far edge, came back into the beams of the headlights, sampled the air before hesitating, and flipped himself around at the top, head raised. He dropped in like a teenage skateboarder on a ramp, straight down the hill, gathering steam before switching back up and around and doing it again. He did that twice more, swooping gracefully, before he dropped down into the flat and ran toward the pile of granite. The scent from the rubble pile had drifted over to the sand dune and crawled up it—Solo was so experienced that he knew what had happened. He'd used those easy swoops to reject the dune as the source of the scent.

Solo only occasionally trained in rubble, and his hot-dog antics on the sand dune worried me. I watched as he stepped up, then up again, on the pieces of granite. I stood well back so he couldn't hear any unconscious gasps on my part as he gingerly climbed the uneven terrain. Mike, standing behind me, said in a low voice, "I tested it."

Solo tested it, too, moving deliberately, sticking his head into the black holes where the scent swirled, moving up, over, back. Then he froze at one hole, turned his head back to stare at me, his eyes glowing amber in the headlights. He slowly backed down off the rubble so he could give his final alert.

"Sweet," Mike said. "Sweet."

18

Wag

I had a dog once. Wag. One of the seven great dogs. At any one time, you know, there are only seven. Did you know that?

— Peter O'Toole as Fisk Senior, *Dean Spanley*, 2008

Solo wasn't the only aging beast in our house. The rest of us were getting sore and creaky; silver crept across the tops of our heads, though Solo's head remained rich red and black. Only his muzzle had grizzled. Megan's entire head was a mixture of white and faded mahogany, her eyes increasingly bleary, as though a fog had descended and was slowly encasing her. She was thirteen, ancient in setter years. Although she was no longer as strikingly beautiful as in her youth, we still used her nickname, Scarlett O'Setter, since she remained as self-centered and spoiled as ever. She continued to demand royalties and obeisance from us. If Solo lay sacked out on a soft dog bed, she would totter over and collapse on top of him, looking reproachful if he startled awake and leaped away from their colliding bones. Her days of tearing my rotator cuff by running out the end of her Flexi lead were gone. Seeing a squirrel would send her into an off-kilter wobble, like a toddler whose attention is diverted. Sometimes she just fell over.

My orthodoxies about dogs and old age softened and shifted with Megan's increasing weakness. We popped mild opiates into her mouth to keep her, and us, happy. We helped her up and down the stairs each night and morning with an elaborate harness that had a rubber handle on top—something I had sworn we wouldn't resort to. We bought her a Martha Stewart quilted dog jacket to keep her warm. She occasionally deigned to gaze on me with approval when I tucked a blanket over her at night. We had Dad's cherry rocker, his good binoculars—and Megan.

Nancy was right: I needed to start another cadaver dog. David and I also wanted to time the puppy's entrance with Megan's departure. We have a small house. Adding another dog in the bedroom at night would assure that it smelled like a bunkhouse. It already sounded like one: Three of us snored. Not Megan, of course. Except for her slovenly drinking habits, she remained a lady, even in decline.

"Why don't you just shoot her?" asked a practically minded K9 officer one night as I explained the Megan dilemma. I stared at him blankly. Dispatching her that way might betray my father's memory, I explained gently. Also, I didn't own a gun. Another K9 officer, Moses Irving, nodded in approval at my answer and glared at his (probably) joking friend. Moses was the minister of a basement congregation in his spare time. "Your father is looking down right now," he said. "You're doing the right thing."

Megan got extra food that night, although she always got extra food if she wanted it. She retained a wasp waist no matter what she ate.

David promised me that as long as the three-dog days and nights didn't stretch out for years, he could put up with the unknown chaos of three generations of dogs.

That promise gave me permission to indulge in what I called puppy porn, scanning dozens of websites and hundreds of photos, raising my endorphins and hopes with two-dimensional images of baby-faced German shepherd puppies. When they are four to six weeks old, shepherd puppies have a flop-eared cuteness that makes everyone go soft and gooey inside. By nine weeks old, they start looking and acting like clumsy tiger sharks. To each her own.

We couldn't get another Solo. I might have been tempted, but Joan Andreasen-Webb was no longer breeding dogs. If I wanted to continue doing cadaver-dog work, my best chances lay with an entirely working-line shepherd. And I wanted to keep going. This time, I wanted to add disaster training to the mix. My former fantasies of a large, calm, red-and-black prince had been replaced with an entirely new fantasy: a sable or black shepherd with a flat back, "environmental hardness," nerve, and drive. I knew exactly what I wanted: one of Kathy Holbert's German shepherd puppies from the mountains of West Virginia, raised with Kathy's mellow laughter and low-key working-dog knowledge, with the gentle hands of her husband, Danny, and with rollicking adventure: crawling through culverts, swimming in creeks, running through the woods, balancing on gently sloped ladders, diving into swimming pools, walking across balance beams. Working-dog heaven.

"You'd better have something for this pup to do once it gets to your house," Lisa Mayhew warned me. She was right. This was not a pup who was going to lie there as Solo now did, snoozing while we watched the new *Sherlock Holmes* on PBS and ate dinner on rickety TV trays. Our morning ritual of coffee in bed with the *New York Times* would be history. Solo slept in and arose, with a luxurious stretch and yawn, only when we decided to.

This would be the kind of pup handlers name Havoc, Harm, or Hecate. We'd kept another name in reserve for years. It had aged nicely and still rolled off our tongues with pleasurable, bisyllabic ease: Coda. Now, though, my goal wasn't a pup who would reflect a quiet, thoughtful ending. I wanted less sonata summation and more Beethoven's coda for his 8th Symphony: fast and furious at points, occasionally disharmonious, "anything but orthodox." This time I would be sobbing in David's arms late at night if the puppy *didn't* immediately leap on us, scrabbling to pepper our arms and legs and noses and toes with puppy bites and claw marks, making us look like heroin addicts. Solo taught me that such behavior wasn't personal, and it wasn't aggression, but rather a gnawing, biting appetite for life. My standards had changed. I knew I could build in obedience, but it was harder to build drive if the basic material wasn't there. I could teach a pup not to leap over the couch and all over us and not to chew on hands. At least, given a few months. Or years.

For all our preparations and research and joy about the choice, I also felt mournful and scared. We were moving into medium-risk territory after several years of relative comfort. A puppy would take lots of time, time away from Solo. That was if Solo even accepted a pup in the house. Also, I would be abbreviating his training in favor of the hot new pup on the block. Getting a new dog up and running could take up to two years—if the pup continued to show promise, if the K9 teams in Durham allowed me to train with them, if no terrible accidents occurred. I had a discussion with another volunteer handler about what might happen if the new dog and I washed out: She condoned finding the dog a new home and moving on quickly to another. I wasn't sure I could do that. I did know our house was too small for three German shepherds. And I was fully aware of the problem of "second-dog syndrome." If the pup didn't work, it would be partly my fault.

I had encouragement in my forebodings. Two experienced law enforcement trainers told me that I would never again have a dog as naturally good as Solo. When I told Nancy that, she scoffed and told me not to be maudlin. "It's the handler, stupid," she said. Within the hour, I overheard Nancy telling a friend that she had just lied to me. I might never again have a dog as good as Solo. I knew that this wasn't the first time she'd lied to me. Her first major lie came when I entered her yard in Zebulon, my belly pack filled with liver treats strapped across my salmon-colored linen pants, hoping to find something for my young *canis horribilis* to do, and she cheerily told me she thought I would love cadaver work. She hadn't thought that. She had seen potential in Solo, but she hadn't thought that I—"little hippie yuppie" that I was—would follow through. Ha. I showed her. Now I was going to have to show her all over again.

"It's very common to see a dog handler be a 'one dog wonder' and to either give it up once that dog is done or to suffer miserably with the next dogs," wrote one cadaver-dog trainer. I exhibited all the clinical symptoms of being a one-dog wonder, and I didn't even have the pup yet.

I was now the working-dog researcher who knew too much. Between genetics and temperament, accidents and poor health, and the limits of my ability as a still-new handler, getting another dog to succeed was a crapshoot. We could stack the odds in our favor, but we ultimately didn't control every contingency.

Maybe I'd been hanging out with law enforcement too much, watching good handlers struggle mightily to understand and respect their new dogs—and failing. Dogs failed, too. I watched almost-adult dogs get shipped in from Europe, get evaluated, and wash out. Not hitting the bite sleeve hard enough. Hesitating before leaping up a metal stairway. Not levitating onto a slippery desk in a warehouse. Mike Baker, who had evaluated many hundreds of dogs, was more patient and knowledgeable than the sometimes judgmental handlers. He knew how long the dog had been in the country, whether it had the equivalent of jet lag, what its early experience might or might not have been. Many dogs faced entirely new environments. Breeding kennels, even top-notch European ones, don't always provide dogs the exposure they need.

And here I would be bringing in a ten-week-old puppy who wouldn't be big enough to climb warehouse stairs. It was bad enough that the pup would be a German shepherd and not a Malinois, but I wasn't bringing in a dozen dogs to evaluate. Only one. So much could go wrong. Solo's gifts were serendipitous and helped shape what he became. There would be no beginner's luck this time around.

On the other hand, I had resources at my disposal. Nancy Hook, for example. Kathy Holbert. Joan. Mike Baker promised me that, though he would be retired from the Durham K9 unit when the pup arrived, he would still be in the K9 training business and would help me put a foundation on the young dog.

During training one night, I tried to assure Mike—and myself—that I wouldn't be as clueless. "I'll know more with the next one. I won't make the same mistakes."

Mike shook his head. He knew better. "If I had every dog in front of me that I'd ever worked with, I'd apologize to each of them."

• • •

Steve Sprouse came to a decision soon after he retired DJ. One last patrol dog.

"My knees say, 'You stupid idiot,' but I just can't picture doing anything else," he told me. Steve wouldn't be getting a nine-week-old pup. "We try to get a dog that's eighteen months, and even at twelve months, we get a dog and there are problems. That six months is really critical." Steve is always looking for two things: genetics and potential.

"I don't want to see what man has put into the dog," he said. "I want to see what God has put into the dog." Steve got lots of eighteen-month-old dogs to evaluate: He would have his pick of a large and mostly unrelated litter from all over Europe. He knew what he wanted—a strong, confident, balanced dog with an internal motor that wouldn't quit. One vendor offered to fly him over to Germany to assess dogs, but soon after that, a vendor he knew well called to say he thought he had a pretty special dog for Steve. The dog had been flown in recently from Slovakia to Florida. Steve didn't take that call for granted; the vendor knew Steve was looking for what would probably be his last patrol dog.

The dog who came out of the kennel toward Steve was no juvenile but a muscled three-year-old, a rich sable male with copper-colored eyes, probably a mix of German shepherd and Malinois, though it was hard to tell. He was "a dog on a mission," Steve said. The dog was looking around, checking out everything, including Steve. Steve did the same with him.

Steve kept the dog's name. Aaron. Hebrew for "mountain of strength."

Despite his age, Aaron was a green mountain. Steve reminds handlers he trains that it takes time and patience and energy to get a dog up to speed, working smoothly, knowing what to do. Now Steve had to adjust his own expectations. DJ had been, as Steve said, "on autopilot." Aaron wasn't. During training searches, he wanted to range out two hundred yards, rather than just fifty yards. He wanted to look around to see if he could figure out what to do without using his nose.

"You have a brand-new dog and you almost expect him to do exactly what the other dog did," Steve said. "He's not sure of the game. That's the hard part. You have to go through that whole process all over again."

Yet Aaron had the potential to be an exceptional patrol dog. Steve knew it. It was just a matter of getting time. Steve was still flying across the country and off to Trinidad to train handlers and dogs, and training handlers in Florida, then coming back to Fort Lauderdale to work with Aaron. It wasn't, Steve admitted, the worst problem to have.

"I guess I may never get to retire and sit in a rocking chair on my front porch, looking over the hills of somewhere in my imagination. It's frustrating sometimes. But I wouldn't trade it for anything else."

. . .

I opened up my file labeled "Very Cool Dog Research" and looked at the studies within to see if any of the researchers had considered the sex of the dog as a factor. Nope.

—Patricia McConnell, 2009

We were standing outside yet another abandoned office building, on a hot North Carolina night. Each patrol dog who had gone into the building, each dog who had come out, had given the K9 salute to the nearby shrubbery. One of the dogs had especially noxious urine; as the dogs coated and recoated the holly bushes, the acrid odor wafted back to us.

The stench inspired the continuing K9 cop conversation about the inherent superiority of male working dogs. I was used to it and could almost ignore it. The law enforcement patrol-dog world is overwhelmingly male and unneutered, creating a self-fulfilling prophecy. Any stray female patrol dog who managed to slip past the gender guards got close scrutiny until she made an error. And she would. Her original error was being female. The next was coming into heat. So when the male dogs got distracted and couldn't work, whose fault was it?

I sighed and rolled my eyes as the stereotypes escalated. Testosterone is undeniably a powerful hormone, its production bumped up by the kind of competition playing out in front of the building. The problem was that I, too, had a preference for male German shepherds: Solo was my third. But I have a feminist contrary streak as well. I talked with Mike Baker, and I talked with Nancy Hook. Mike pointed out that one of the toughest dogs on the K9 team had been a female. Females tended to be less distractible than males, he said. Nancy said it depended on the dog, and I should get the pup I liked. She'd approve of any female as long as it was a "bitch from hell." Nancy would make sure I didn't put too much obedience in the dog, so she didn't stare adoringly at me and ignore her work. I thought about the cadaver and tracking dogs I had been reporting on—from Lisa Higgins's to Kathy

Holbert's to Andy Rebmann's to Marcia Koenig's to Jim Suffolk's to Roy and Suzie Ferguson's. A mix of sexes, with just as many females as males. Roger Titus, bless him, had used all female bloodhounds but one in his tracking career. That male was an early mistake, Roger said.

Joan helped nudge me further along the female trail. "I love the way females work," she wrote me. "Totally different kind of relationship . . . at least, for me. I always found it interesting that Stephanitz also preferred the work ethic of females."

Max von Stephanitz—the most sexist of them all, the man who explained that German shepherds would obey the woman of the house only "with reservations"—preferred the work ethic of female shepherds?

Neither Stephanitz, Mike, Nancy, nor Joan tipped the balance. Solo put his big paw on it. He had grown up a lot in eight years, but I could easily imagine a male adolescent shepherd telling him to shove it when signals got crossed one too many times. One doctoral dissertation showed that more egalitarian relationships and play occur between male and female canines than between same-sex pairs. That made sense. Even socially challenged Solo would ultimately start to show off for female dogs, cavorting goofily, rather than bristling. That was my hope: more romp and less hackle.

How, though, could I assess which female puppy to get?

Of course, I fell back on research. I found a great deal of work on "whorls"—those mathematically inevitable spots where hairs converge and then wheel one way or the other. Cowlicks. Temple Grandin's early work in cattle-hair whorls showed that the direction and position of the hair swirls on a steer's forehead helped predict whether he was calm or fearful.

Australian veterinary researcher Lisa Tomkins went to town based on that work. She assessed 115 future guide dogs, looking at their whorls and their paw preferences. Then she followed their progress. Puppies who preferred to use their right paws over their left were twice as likely to pass guide-dog school. Puppies with counterclockwise chest-fur whorls were more than twice as likely to succeed than those with clockwise chest whorls. Tomkins and her fellow researchers noted that it appeared to be linked with the whole left-brain/right-brain crossover. I already knew Solo preferred to use his right paw to snake toys out from under the sideboard if he couldn't use his mouth. I looked at his chest with some trepidation; at first all I could see was an undifferentiated mass of fur, but as I parted the fur and moved farther down his chest, just before he leaped up, frustrated by my nonsense, I saw one small cowlick moving in reverse. Thank God.

I could hardly wait to tell Mike and Steve how they could save police K9 units time and money on shipping dogs in from Europe to assess. As a first order of business, all they had to do was obtain full frontals of potential patrol dogs' chests and hope that the brokers didn't know enough about the research to flip some photos to ensure that all the dogs had counterclockwise patterns. Oh, and Mike and Steve could ask the brokers if they wouldn't mind giving the dogs Kongs filled with frozen treats to see which paw a dog used to hold down the Kong and remove the treats. It was even possible that a female or two might pass that test.

I could envision my and David's arrival at Kathy Holbert's hilltop kennel. I would have my checklist in one hand, a food-stuffed Kong in the other. Here's what we need: one female German shepherd puppy with a counterclockwise chest whorl and another ideally on her right elbow, and a strong right-paw preference. Along with everything else: high drive, high sociability, great health. And a sense of humor. Hope that's not too tall an order?

. . .

The pups were born in early September, before the leaves of the huge sycamores started to turn yellow in the mountains of Barbour County, West Virginia. Their father was such a dark sable that anyone would have called him black, though he had traces of dark velvet brown on his chest and belly. My lookism returned momentarily, although I tried to tamp it down. He was stunning. I had already fallen for the pup's mother, Kathy's trailing and article-search dog, Reza. She made everything look easy. She'd toted around toys until she realized she was in labor: Whoops. Excuse me. Got to deliver some pups. She would go back to the toys soon enough, bringing them to the pups to play with. She fell into the category of "fun mom."

I watched the three female pups emerge on Facebook, via messages and photos, with Danny holding up each one for Kathy's camera, three moles getting their first mug shot. One had a drifting trace of umbilical cord still attached. She held both pink paws up, almost as though trying to protect her homely squinty face from the camera's unforgiving lens. I couldn't tell if she was pushing her right paw out more than her left. Two females were behemoths right out of the womb, both more than a pound. The third female, black like one of her sisters, was half their size. Nine ounces. Kathy named her "Little Bit."

We assume puppies learn all sorts of things from their mothers. But do they learn by watching and then imitating, or is it instinctive, like a mama cat leading a kitten to the litter box or a mare leading a colt to water? Nancy suggested that Solo could teach the pup a thing or two. Of course, not all teaching is good teaching. I'm pretty sure Megan's raucous behavior encouraged Solo to bark when the UPS man arrived, but she could simply be infecting the air with bad-dog vibes, ramping up Solo's arousal level.

Older dogs aren't immune from teaching. Steve Sprouse discovered that Aaron had a few issues with water: He hated the hose, the sprinkler, and the swimming pool. His life as a stud dog clearly hadn't been halcyon. Probably someone had used water to discipline him, or to separate him from females after breeding. But patrol dogs need to tolerate water in Florida; they're surrounded by it. Steve watched Aaron watch Casey, their female shepherd, run into the sprinkler after her toys. Aaron liked Casey a lot. Soon Aaron was diving after his toys as well. Though it was a good first step, it wasn't all wonderful emulation.

DJ has a bad habit of spinning in his kennel when he's hyper. Now Aaron spins in his kennel.

Most cognitive psychologists would ascribe Aaron's behavior to something other than imitative behavior. To mimic behavior after witnessing it was thought to be distinctly human. While we call it "aping" or "copycat" in a denigrating manner, watching someone do something and then trying to do it yourself isn't a low-level cognitive act. It's part of what makes the machinery of human culture chug along. The old joke about a new bride's brisket applies: Sylvia makes her mother's perfect brisket recipe by cutting off the ends of the roast, the way her mother always did. She asks her mother why, and her mother tells her that's the way her mother did it. Sylvia finally asks her grandmother. Because, her grandmother informs her, the pan she has is so small, the only way the brisket fits is with its ends cut off. This is what it means to be human.

Ascribing imitation only to humans is starting to change, one peer-reviewed article at a time. Dogs aren't leading the experimental way, but they are part of a diverse animal pack, joining corvids, meerkats, marmosets, and elephants. Little research has been done on "social learning" or "imitation" or "observational learning" in working dogs, though that, too, is starting to change. As Deak Helton already pointed out, it's harder for scientists to get access to working dogs. Only one small published research study shows what many handlers believe is the case—working dogs learn from observing other dogs. The 1999 study from South Africa used two litters of German shepherd pups: One litter got to watch their mother, a drug-detection dog, at her job. That's all they were allowed to do—watch but not participate. The other litter didn't get to see their mother work. When the pups were six months old, 85 percent of the pups who had observed their mother at work passed an aptitude test for narcotics. Fewer than 20 percent of the non-observing pups passed.

More good news for Solo and the new pup was that dominant animals generally do better than submissive ones at being good demonstrators, solving problems, and learning new tasks. I'm not saying that, just because it works for lemurs and chimps and domestic hens and mice, it should work for German shepherds, but that's not such a leap. Decades ago, my brother Mark and I watched two dogs at the beach. One of Dad's Irish setters couldn't climb a steep sandy cliff. She'd tried and fallen. Tarn, my first German shepherd, ran up the cliff and then stood and stared down at her. She whined. Tarn came down and went back up, slowly this time. Dad's setter watched, tried to follow, and failed. Tarn came down a third time, went to her heels, and barked loudly. She shot up the cliff. It's just an anecdote. It might only demonstrate Tarn's instinctive herding behavior. I choose to believe both that Tarn was demonstrating what to do, and that Dad's dog was doing her best to imitate him. Or avoid him.

David and I watched the videos that Kathy posted of the four puppies—three females and a male—as they started negotiating the PVC-pipe ladders, culverts, and barrels she had set up for them in the play yard. We tried not to fall in love with any pup in particular. Videos don't tell the story; nor do photos or even e-mail. That would be like picking a pup off Dogmatch.com: fantastical and engaging and perfect without having to make a commitment or know more.

The two big females—one black, the other black and tan and similar to her mother in looks—showed signs of handsome adulthood. I avoided gazing directly at Little Bit's face. With her wide milky-blue eyes, midnight fur, and chunky yet delicate nose, she looked like a plush Japanese anime puppy. I wasn't afraid she would turn me into stone but into mush.

I also had been small and cute; people had patted me on the head a great deal because they could. I didn't have the evolutionary advantage of sharp puppy teeth to keep them from the top of my head. Little Bit brought out the protective "aww" in people, not the awe I ultimately wanted.

When their mother wasn't there, one or two of the puppies would try something, and then another one would try. Let's go climbing.

Let's break out of this barrel. Let's slip and slide through this metal culvert and bite each other. I'm not a cognitive psychologist. All I knew, watching the videos, was that their constant play and experimentation together, even what appeared to be occasional, perhaps accidental cooperation, was additive and exponential. They were raggedy, impulsive, enthusiastic, distractible. Their coordination wasn't quite as smooth as a rugby team's, but you could see hints of what their future might hold: an entire pack of puppies, tails wagging, diving into dense underbrush and climbing over rubble to search.

At first Little Bit couldn't climb over obstacles as easily as the larger pups. In one video, Kathy chuckled quietly as five-week-old Little Bit growled in frustration, flinging herself repeatedly at a broad plywood teeter-totter, a wobble board, until she finally scrabbled on, using her back feet and belly to propel herself like a turtle. Then she ran off to play games by herself, sliding around in a big metal culvert, grabbing her own tail, going in circles inside the circle. I showed David the videos of her and told him sternly not to get attached and to look at all the strengths of the two big handsome females. It had been easier with Solo; there hadn't been a choice.

In the end, Kathy made the choice for us. She called me when the pups were eight weeks old. She had spent all day evaluating them—one of the big females I had yearned for early on was "neck and neck" with Little Bit in hunting tests. By the end of the day, it was clear to Kathy that Little Bit hunted longer and harder for her toy than any of the other pups. It was behavior that one needed in a cadaver dog. Little Bit would disappear in the dark down the road, worrying Kathy, and then come back toting a lost ball in her mouth. She found her mother's blue ball in the snow before her mother did, and grabbed it, the whites of her eyes showing against her black fur as her mother turned sideways to try to snatch it. No dice. Little Bit was independent and contrary. She would be a pain in the butt and a joy to train. A plush toy with razor-sharp teeth and a brain.

David smiled broadly when I got off the phone. Early evaluation isn't destiny, but it helped our confirmation bias.

Kathy told me that she and Danny had a five-minute "pity party," as Kathy called it. This pup had stubbornly squiggled her way into their hearts. Then Kathy stopped calling her Little Bit and started calling her Coda.

. . .

We pushed hard on the toll roads, driving from West Virginia back to North Carolina to beat the setting November sun. I stewed most of the way home, certain Solo would kill our precious sleeping cargo with one big paw and a bite. So much for Little Bit. I went back and forth on arrangements, micromanaging the details of their meeting, dreading the tragic outcome. We decided that I would drop David at the house, he'd exercise Solo thoroughly, and then he'd bring him to an empty ball field a mile from the house, where Coda and I would meet them. Megan was on an extended playdate with our good friend Barb Smalley and her dog, to keep introductions simple.

We were punchy with exhaustion, sore, hungry—too old for this nonsense. Dusk was settling in on the ball field and surrounding woods as I opened the back door of the Camry. I left Coda in her crate and stepped away. Solo leaped from the Civic, greeted me exuberantly, and went over to sniff boulders and lift his leg. I had to call him over to the Camry. He briefly sniffed Coda through the wire grate, then returned to the more enticing urine on the rocks. Obviously not car possessive. While he was distracted, I carried tiny Coda out onto the field, set her down, and backed a few steps away. David had Solo's red ball. I had treats. Coda barked sharply five or six times at the approaching monster, then ran back to stand between my legs. I heartlessly backed up a bit more. Solo sniffed her, hackles up, tail high and wagging, then low and wagging. The hackles came down, and he opened his mouth and grinned. He stretched his big body nonchalantly, luxuriously, over the black puppy, whose outline was starting to blur into the dusk. She sat up underneath him, barely skimming his belly, one ear flopped over like a bang, one already straight and cone-shaped. Solo flipped toward David, levitating his back legs in a delicate pas de deux, leaving Coda untrammeled. She no longer looked cowed, but curious. Solo stared at David. You're the one with the ball. So let's play.

Acknowledgments

Solo won't appreciate being acknowledged, but he is the genesis of this book. For several years, I rejected the idea of writing about cadaver-dog work, although I had done one brief feature story early in Solo's training. I kept insisting, when friends asked, that I was too busy living it, that the act of capturing the work in words would actually ruin the joy, steal its soul, and fuse two worlds I wanted to keep separate. That was the closest to superstition I ever got, except when I had to fly. I worried I might inadvertently expose secrets of the working-dog temple I'd been allowed to enter, and alienate the people I respected. But I kept falling further in love with Solo, the work we were doing, and the people we trained with. I realized, despite my protestations, that I wanted to capture Solo in all his scary, funky, funny uniqueness, and remember critical moments of the wild new ride he had taken me on. When Solo turned six, I turned to David and said, "I want to write about him." It was almost that simple. I love first steps. I also don't want to forget. Inevitably, one does.

This is a book about people and dogs, which includes many who aren't dog people as well as those who are. Over the past three years, since I started this project, a multitude of people have helped me create this book: dog people, scientists, law enforcement people, medical examiners, and many close friends and colleagues.

First, I want to acknowledge a group of people I cannot name because they are linked to cases: the dedicated, patient, knowledgeable, and brave members of law enforcement and search specialists across North Carolina, the ones willing to go down in the swamp, to bushwhack, to think through all the possibilities. The work they do is grindingly difficult, and I cannot thank them enough—for the safety they provide, for the care they bring to their work, and for what they have taught me.

Then comes the dog-people triumvirate: Joan Andreasen-Webb, Nancy Hook, and Mike Baker. Joan, Solo's breeder, taught me about German shepherds for the first time in my life. Her patience and kindness and deep knowledge of dogs, especially the German shepherd, continue to inspire me. Nancy, Solo's and my first and continuing trainer, rearranged my priorities and gave me a new way to be with dogs. And I have been enormously fortunate to train with Mike. His enormous talent, his deep knowledge of dogs, his quiet authority, his unending patience, and his ability to make dog work into dog play are an inspiration.

A fourth person, Lisa Mayhew, is part of that early and continuing circle of influence and friendship. She introduced me to Mike and to other working-dog people across the state. She helped train me and Solo, and especially helped me understand and navigate the complex intersecting worlds of forensics, dogs, and law enforcement.

Other North Carolina K9 trainers and handlers tolerated me and sometimes welcomed me, helped me train Solo, and let me learn by watching them. Whether or not they are directly named in the book, they taught me a great deal. Both the Durham Police K9 unit and the Durham Sheriff K9 unit have allowed me to come to their weekly and monthly trainings for years. I want to especially acknowledge now-retired Durham Sheriff K9 Sergeant Rick Keller and current Durham Sheriff K9 Sergeant Steve Talley. I was fortunate to train and work with

Durham Sheriff Deputy Tim Fields when he had a cross-trained cadaver dog, and currently with Durham Sheriff Deputy Brad Kirby. On the Durham Police K9 unit, I want to especially thank Danny Gooch, Cindi Wood, Moses Irving, Terry Tanner, and Kristy Roberts; many of them are retired or have moved to new positions in law enforcement, but they all taught me a great deal. Alamance County Sheriff K9 Officer Kathy Edmonds has been wonderful over the years. Both the Reidsville and Gibsonville police departments have welcomed me at their trainings, and I am grateful for the continued ability to learn from them. Ken Young, Darlene Griffin, and other members of the Triad Bloodhound Team nurtured my early passion for this work.

Next come those K9 trainers and handlers whom I learned from at conferences and seminars and who taught me more about handling Solo at the same time I was researching and writing.

First and foremost are Andy Rebmann and Marcia Koenig. I canceled one planned trip to Seattle, with Marcia explaining apologetically that Andy had snapped and dislocated some parts, including an Achilles tendon and a shoulder, and they were having a bit of trouble stabilizing him, so it might be better to schedule a later interview. Andy pulled through, and it was worth the wait. Their generosity, humor, deep and broad knowledge, and passion for the work inspire me almost as much as Andy's creative profanity does. And it was Marcia who suggested I contact Kathy Holbert of West Virginia, who became not only a critical part of this book but of our lives. She is a wise and wonderful person and trainer, and I've already learned so much from her and from the next generation she gave me and David: Coda, our funny black trout of a German shepherd puppy.

Re-creating the early days of dog research was not an easy task, and it was at times poignant, as a number of the central figures in this early work are deceased or in declining health. My sincere thanks go to Jim and Sally Suffolk, Nick Montanarelli, Roger Titus, Deborah Palman, Ed David, Jim Polanis, and Joanna Johnston for their help, their good records, and their fine memories.

ACKNOWLEDGMENTS

I had the good fortune to train and observe at various seminars and trainings across the country researching this book. Broward Sheriff patrol K9 trainer Steve Sprouse welcomed me into his training seminars in both Florida and North Carolina. His wife, Sandy, welcomed me into their home and served a mean lasagna. Like Mike Baker, Steve has a plethora of handling skills and dog and people knowledge that leaves me awestruck and appreciative. Thanking Steve for his significant contribution to this book feels inadequate.

At cadaver-dog seminars in Georgia, Mississippi, and North Carolina, I was able to observe and train with Lisa Higgins, one of the best teachers and mentors I've ever had. And a special thanks to Lisa's granddaughter Haylee, who helped keep things warm and funny and real at dog trainings, and who reminded us handlers what the work is really about. Roy and Suzie Ferguson of Tennessee welcomed me into their home and to their trainings, and introduced me to their mentor, Art Wolff, and his brother, Ed Wolff, also an experienced K9 handler. Their combination of openness, professionalism, and joy in the work is a model I aspire to. Mary Cablk and her husband, John Sagebiel, hosted me and David for a wonderful long weekend in Reno, where I watched Lisa Lit train with Chris Salisbury and others, and talked with both Chris and Cindee Valentin. I am so grateful for the thought, care, and attention that Mary and John gave to us—as well as the wonderful food, thought-provoking conversation, and continuing help.

At the same seminar where I joined Andy and Marcia, I was able to spend time with Kevin George, one of the most talented and imaginative trainers I've ever met. Besides knowing magic, his story-telling abilities made me laugh until tears ran down my face. Nancy was downright jealous when she heard I attended one of his seminars and got to be his assistant.

Roger Titus, in addition to helping with the history section, shared his deep knowledge of bloodhounds and all trailing and tracking dogs with me on a beautiful fall day in the Triad of North Carolina. If you run bloodhounds, I think that humor, patience, and tolerance must run in your genes. Roger had no reason to be patient with me, but he was. His continued generosity means a great deal to me.

Paul Martin, who helped found and still helps run the Western Carolina Cadaver Dog workshops, is another one of those trainer/handlers who went above and beyond in his work with me. His sly humor, low-key approach, and willingness to share his knowledge and his developing research have been crucial to this project.

Brad Dennis and Deborah Palman helped in the evolution of my thinking about searches, and I'm grateful for the time with them. With their decades of experience and knowledge of both wild and domestic human and animal behavior, we are fortunate they are part of the search world. Matt Zarrella was enormously generous with his time on the telephone sharing his dog work and insights. Suzi Goodhope was the equivalent of a (just slightly) older sister at seminars, making me feel as comfortable and warm as though I were wearing one of her knitted caps with the skull and crossbones on them. And Charm Gentry was the equivalent of a matchmaker, laughing at me and with me, and letting Kathy Holbert know I wanted a puppy from her kennel more than anything.

I also want to acknowledge a long list of handlers and trainers I was able to witness work and to interview both formally and informally. I am so appreciative of what I learned from all of them. Even if their names are not explicitly used in the book, they expanded my knowledge base immensely: Michael Ben Alexander, Orval Banks, Kathi Brown, Shelly Burton, Ann Christensen, Maria Claxton, Terry Crooks, Paul Dostie, Melissa Ellis, Melissa Frye, Karen Gillespie, Gwen Hancock, Deana Hudgins, Nikki Ivey, Renae Johnson, Sean Kelly, Lisa Lepsch, Dave Lopez, Roxye Marshall, Joe Mayers, Paula McCollum, Helen Moreno, Marshia Morton, Benjamen Ortiz, Craig Patton, Brooke Proctor, Pete Sepot, Becky Shropshire, Meaghan Thacker, and Sue C. Wolff.

Several organizations and seminars and police departments across North Carolina and the country allowed me full access to observe trainings, take photographs, and sometimes participate with Solo. I am so grateful to the National Search Dog Alliance and Jan Meyer; DogMeet 2011 in Washington and Bruce Ramey; the Network of Canine Detection Services, especially Ronda Maine, Barbara and Danny Holley, and T. H. Walker; Western Carolina University Cadaver Dog Workshops, especially Bobby Hensley; the Tennessee Special Response Team-A; the Greensboro Police Department K9 unit; Brigit and John Hall of Alert K9; and Southern Pines Police K9 unit.

This book also depended greatly on many scientists, medical examiners, epidemiologists, veterinarians, dog behaviorists, archaeologists, anthropologists, historians, amateur historians, military personnel, and environmental scientists who were unstinting in their willingness to e-mail, talk, and fact check. Each person on this list has my deep gratitude and thanks: dog-training book author Carol Lea Benjamin, former *American Scientist* editor Chris Brodie, archaeologist Haleh Brooks, Wendi Dick, Pat Franklin, Ken Furton, epidemiologist David Goldsmith, Deak Helton, Western Carolina University forensic anthropologist Cheryl A. Johnston, Lisa Lit, May MacCallum, Patricia McConnell, Kha McDonald, Charlie Mesloh, Larry Myers, Raincoast Conservation Foundation's senior scientist Paul C. Paquet, southeast ecologist Milo Pyne, Ephraim Rotter, John Sagebiel, Greg Sanson, Solo and Megan's veterinarian Robin Scott, veterinarian Tami Shearer, Michael Sledge, Marci Sorg, neurologist and bear brain expert George Stevenson, Mississippi archaeologist John M. Sullivan, botanist Wade Wall, University of Washington's Center for Conservation biology director Sam Wasser, Barbara Weakley-Jones, Western Carolina University forensic anthropologist John Williams, and friend and epidemiologist Steve Wing. A special thanks to Arpad Vass for his time and generosity.

I don't know where to put Nancy's daughter Lindsay and Nancy's grandson Sean in the mix. They cross so many categories, helping train me and Solo from the time Lindsay was a teenager and through Sean's birth to the present. Sean is simply a fine, fun kid to have around: smart, tough, funny, and tolerant of people and dogs.

The academic community has a few proponents of dog books. Thanks to Cary Nelson and Paul Treichler—partners, Samoyed lovers, and my academic mentors—for being amused, impressed, and incredibly supportive when they learned I planned to both write a dog book and be the editor of a national magazine for the American Association of University Professors while Cary was still president. Although in that moment when Cary asked me why I seemed to be having trouble writing a straightforward book about a dog, I didn't have an answer. It was a very Cary moment. He was right, so I shut up and kept writing. Thanks to Elaine Orr for her quiet, fine advice along the way. Tony Harrison, my department head, has fully adjusted to the strange reality of my project and has been enormously supportive in ways large and small. I am so appreciative of being in a department that houses a variety of disciplines.

This is my first single-authored book. I developed the proposal with the help of Laurel Goldman and the members of my Thursday morning writing group in Chapel Hill: Anna Jean Mayhew, Fabienne Worth, John Manuel, Betty Palmerton, Eve Rizzo, Mia Bray, and Cindy Paris. They were tough and kind and supportive. Reading aloud has never been more terrifying or rewarding. Laurel's exactitude and talents as an editor humble me. And thanks to the Weymouth Center for the Arts & Humanities, which gave me four days of peace to finalize the proposal, although I was deeply distracted by the Friesian horse stables next to the old mansion and by the deep bass of the bullfrogs that inhabited the ponds in the garden. Thanks, too, to Jim Roberts and Nick Roberts of Doe Branch, Ink, a sylvan mountain writers' retreat where I worked on the proposal with Peggy Payne, and another wonderful group of wise and thoughtful listeners: Mahan Siler, Susan Schild, and Fabienne Worth. Using a faint mountain Internet connection, Peggy showed me how to find an agent through Publishers Marketplace; that introduction led me, almost immediately, to finding Gillian MacKenzie and her book projects, which I loved. Gillian didn't know me, and I didn't have a single name I could drop in sending her my proposal via e-mail one cold Sunday night in January. Yet, in less than twenty-four hours, I had the warmest, most comprehensive, and insightful response from her. That level of response hasn't changed; Gillian helped expand and shape the proposal and, because of that, the book, with a sure and talented touch. Then she passed the baton on to Michelle Howry, senior editor at Touchstone, and I am so grateful for her warm, smart editing, her thought and care, and her continuing attention. I have a number of writer and editor friends, so I know how lucky I am to have both Michelle and Gillian lending their enormous talents to this project. My thanks to Michelle's editorial assistant, Brendan Culliton, and to Anna Jean Mayhew, who helped edit at critical points in this never-ending book process. The team at Touchstone, Jessica Roth, Meredith Vilarello, and Linda Sawicki, have been a joy to work with. Many thanks to David Falk.

Thanks go to photographer D. L. Anderson, who contributed hugely to this project with his keen eye and his ability, thank goodness, to avoid anthropomorphizing dogs; to Sherri Clendenin, whose early photos of Solo helped me fall in love with him; to Steve Sprouse for his considerable talents as a patrol dog photographer; and to Nick Montanarelli for his early snapshots of dog research. My thanks also to Lissa Gotwals for her excellent author photographs. Seth Mulliken, Robert Bell, and Brooke Darah Shuman all played central roles in shooting video of Solo training. I thank them all for the memories, as well.

My friends, my and David's friends, and our families became collaborators and played critical roles along the way. Dale Maharidge helped consume a fair amount of wine and food one night in Manhattan, and outlined all the reasons I could do this book and why it would be the beginning of the rest of my life. We have known each other for thirty years, and Dale has not only helped me every single step of the way, he's often anticipated what I need to do next in this complex process. I could not have a more wonderful cheerleader, friend, and coach. Sherryl Kleinman, who has never had a dog but has fantasized about the perfect, low-maintenance dog, helped sketch out the chapters on long drives across North Carolina and back and forth from Washing-

ACKNOWLEDGMENTS

ton, D.C. I'm devastated that the chapter titled "Sherryl Wants a Dog" didn't survive the brutal editing process. Friend and colleague Sharon Setzer teased and cajoled and supported me during the writing of "my pet memoir." Sarah Stein listened to my stories about training and searches, has the kind of faith I don't, encouraged my soul-searching, and took incredible video of Solo training. She was my spiritual guide for this book journey. Scott Browning laughed at my fear of writing, read with great care, and sent me news articles before they existed online. Ronnie Cohen, whom I've known even longer than I've known Dale, edited and coached with such enormous skill and care that I don't know what I would have done without her. Barb Smalley took our sweet Megan for overnight "playdates" so I could work. Our neighbors Michael Hardt and Kathi Weeks, cat people who nonetheless like our dogs, took care of us with simple bottles of wine and good conversation in the backyard—impromptu gatherings that demanded little and gave much. Kathi introduced me to Kathy Rudy, a dog person, who has become a mentor and supporter, helping me think through the project. Our other neighbors, Bruce and Dorean Sanfelici, let the dogs out and let me let my hair down. Anita Levy isn't a dog person but cared anyway. Kim Turk, Leila May and Don Palmer, Jennifer Washburn, Barry Yeoman and Richard Ziglar, Mike Ferguson, Diane Johnson, Anne Eckman, and David Schulman listened and helped and commiserated every step of the way, from a distance and from close up. And Deborah Hooker listened and cared and helped me articulate emotions that I'm not sure I knew about before I told her. My high school friend and locker partner, Randhi Sue Wilson, introduced me to German shepherds with Tolkien, her shepherd mix who went everywhere and showed me that dogs were meant to be partners. I thank them all for their friendship and help.

Hal and Patsy Hopfenberg helped us escape to their mountain home with Solo, where we ate, drank, talked—and I wrote. And when I was depressed and overwhelmed and frantic, and David was tired of dealing with me, we would drive the 1.1 miles down the road and get

289

a little table at the bar of the Magnolia Grill. Ben Barker, chef–owner, would take a brief break from the kitchen and come over to chat, even though he knew our arrival meant I was in a foul mood. I would whine, taste Ben's warm brown food, and feel so much better that by the time I ate one of Karen's amazing desserts, I would have forgotten what ailed me. The Grill happened to have the best food and desserts in the country until it closed last year after twenty-five years, and I am everlastingly grateful for its existence. We will soldier on, despite its closure. Thanks to Ben and Karen and our friends Joe Levine and Louise Antony, David and I got engaged at the Grill. They helped sustain this book—and our marriage.

David's siblings in New York, Bob and Irene, and their respective spouses, Arleen Auerbach and Phil Schaeffer, were amazing—I thank them for being such wonderful in-laws to me, and for their understanding that our schedule went a bit south with this book project. We will visit much more in the coming years. My brother Mark talked through early memories of Irish setters, my nephew Kelly loved the very idea of this book, and my brother Dan, who has hunting Labradors, understood my predilections for working dogs.

My father, Charles Warren, who died of cancer in 2005, left me, at the supposedly mature age of forty-nine, in the deepest mourning I have ever experienced. David lost both his parents when he was a teenager; I lost my mother more than a decade before Dad died. But his death still feels near. I can still hear his slow, thoughtful voice. I still want to pick up the phone so that we can talk through the state of the world and our lives. He is woven throughout my life and this book. I am so grateful to my dear stepmother, Agnes Rands-Warren, for making Dad's last years so happy and being a "best-mother" to me.

Finally, there's David: my husband, my heart, my dearest friend. I don't know how to acknowledge him, except to say that I love him, and I promise not to do this again too soon.

Photo Credits

Cover photograph of Solo, seven and a half years old.
(Photo by D. L. Anderson)

1: The Little Prince of Darkness (p. 1)
Solo, four weeks old, at breeder Joan Andreasen-Webb's home in Pataskala, Ohio.
(Photo by Sherri Clendenin)

2: Death and the Dog (p. 15)
A statue of Anubis, the Egyptian jackal-headed dog god who protected the deceased in their tombs.
(Photo by Son of Groucho)

3: Nose Knowledge (p. 27)
Roger Titus, vice president of the National Police Bloodhound Association, rewards a bloodhound for finding him during training.
(Photo by Cat Warren)

4: Birth of the Body Dog (p. 47)
Military and Southwest Research Institute studies showed the value of using dogs. The snapshots of dogs training during the 1960s and 1970s came from researcher Nick Montanarelli.
(Photo montage by David Auerbach)

5: The Shell Game (p. 69)

Durham K9 Sergeant Mike Baker and K9 Officer Danny Gooch work with Danny's young patrol dog, Rin, using boxes where only one has a drug scent.

(Photo by D. L. Anderson)

6: Distillations (p. 83)

Human remains can easily disappear in the woods, becoming indistinguishable from the surrounding flora.

(Photo by D. L. Anderson)

7: A Spare Rib (p. 99)

Solo created new training problems for me. Nancy Hook laughed and called him a jackass.

(Photo by D. L. Anderson)

8: Comfort Me with Bite Work (p. 109)

Broward County Sheriff K9 officer-in-training Dave Lopez takes a bite from Diesel, a German shepherd learning to work in the water.

(Photo by Steve Sprouse)

9: Into the Swamp (p. 131)

Solo running through undergrowth in the woods during training.

(Photo by D. L. Anderson)

10: Cleverness And Credulity (p. 145)

Solo is smart and devoted, which means he wants to please and to get his reward. That can be a problem.

(Photo by D. L. Anderson)

11: All the World's a Scenario (p. 161)

Setting up realistic scenarios during training is crucial to both the dog's and the handler's success.

(Photo by D. L. Anderson)

12: The Grief of Others (p. 175)

Andy Rebmann and his German shepherd Josie searched for days for the victims of a serial killer or killers along the highways near New Bedford, Massachusetts.

(Photos by Paula Bronstein, *Hartford Courant* article by Lynne Tuohy)

13: All the Soldiers Gone (p. 189)

In Iraq, German shepherd cadaver dog Strega alerts on a training aid hidden by her owner and handler Kathy Holbert, a civilian contractor from Philippi, West Virginia.

(Photo by Army Staff Sergeant Daniel Yarnall)

14: Running on Water (p. 201)

In Mississippi, Paul Martin helps Gwen Hancock train her Labrador, Ruger, on human remains detection in the water while Cathi Brown observes.

(Photo by Cat Warren)

15: The Perfect Tool (p. 217)

(Photo by D. L. Anderson)

16: Grave Work (p. 231)

Lisa Higgins and her cadaver dog Maggie train at the Cobb family cemetery in Tuckaseegee, North Carolina, where research that combines family history, cadaver dogs, ground-penetrating radar, and other methods is being used to locate grave shafts and unmarked graves.

(Photo by Cat Warren)

17: A Second Wind (p. 249)

Sean Kelly, a K9 officer in North Carolina, during a training break with Nero, a former military working dog.

(Photo by Cat Warren)

18: Wag (p. 265)

Solo and Coda at home in Durham, one week after Coda's arrival.

(Photo by D. L. Anderson)

Notes

The source materials for this book include my own training experiences, especially since I started working with Solo in cadaver work in 2004; my observations and participation at numerous search-and-rescue and cadaver-dog seminars and at police and sheriff K9 trainings and seminars; and my experience on searches and deployments. The book depends on personal communications, off-the-record conversations, on-the-record interviews, and lengthy correspondence with dozens of dog handlers and trainers, members of law enforcement, search-and-rescue volunteers and managers, forensic anthropologists, archaeologists, botanists, analytical chemists, cognitive scientists, epidemiologists, veterinarians, conservation biologists, medical examiners, military researchers, and historians. In addition, I am grateful for access to the personal archives and training and deployment records of several experienced trainers and handlers. I also depended on literally thousands of newspaper and magazine articles, academic articles and conference proceedings, military reports and studies, and several dozen books.

Oliver Sacks noted in a February 2013 *New York Review of Books* essay, "There is, it seems, no mechanism in the mind or the brain for ensuring the truth, or at least the veridical character, of our recollections." I agree. However, when I quote someone in direct quotes, those come from extensive notes, tape-recorded conversations, e-mails, or other correspondence. In a few instances, the words were seared into my brain because I'd heard them often and repeated them to my husband, David Auerbach, and to many others. Durham Police K9 Sergeant Mike Baker's standard advice to handlers, "Be more exciting than pee on a tree," is just one example.

These endnotes, like the book itself, are not comprehensive and are not meant to be, but represent highlights of the materials used. In some cases, I don't use people's names because I was an outside observer at seminars or trainings where handlers and trainers graciously allowed me to take notes and photographs. Although I don't use names in every instance, I am deeply grateful for the knowledge imparted. I did not use composite characters or composite quotes, nor did I change names or details. In several instances, I don't identify certain particulars about a search, but I did not in any other way change the facts of any case.

1: The Little Prince of Darkness

Interviews and correspondence for this chapter include Solo's breeder, Joan Andreasen-Webb; my husband, David Auerbach (who is present in every chapter in some fashion); Nancy Hook of Hook's K9 Training; and our friend Barb Smalley. Joan exchanged dozens of e-mails with me in those early Solo days, and later noted that she wished she could have lived closer, as she felt somewhat helpless from a distance.

The German shepherd is a controversial breed (many breeds are). For the German shepherd fan, no argument is as heated as the one about whether Americans, especially the American Kennel Club—with its insistence on a distinct and standardized look for particular breeds over other qualities, such as health and working ability—ruined the German shepherd. There is also a "purer" history of the German shepherd that working shepherd lovers tend to equivocate about or simply avoid. That history began with Max Emil Friedrich von Stephanitz (December 30, 1864, to April 22, 1936), a German army officer who developed the German shepherd breed as we know it today; set guidelines for the standard; and was the first president of the S.V., which stands for *Verein für Deutsche Schäferhunde,* or Society for the German Shepherd Dog. His book, *The German Shepherd Dog in Word and Picture* (Jena, Germany: Anton Kämpfe, 1925), weighs in at 710 pages. As Susan Orlean notes with light irony in her book *Rin Tin Tin: The Life and the Legend* (New York: Simon & Schuster, 2011), Stephanitz had a "Germanic enthusiasm for genetics" (23). Nonetheless, he reportedly only reluctantly turned his registry over to the Nazis, who took up German shepherd breeding with alacrity. Two helpful resources on the topic include *Animals in the Third Reich: Pets, Scapegoats, and the Holocaust,* by Boria Sax (New York: Continuum International Publishing, 2000), and "Breeding Racism: The Imperial Battlefields of the German Shepherd Dog," by Aaron Skabelund, *Society and Animals* 16, no. 4, 2008: 354–371. Working-dog aficionado James R. Engel also provides a great deal of history and context in his web-only book project, *The Police Dog: Evolution, History and Service,* http://www .angelplace.net/Book/.

The history of the dog sport of Schutzhund is long and complex. It was designed at first to help measure working dogs' abilities for breeding programs. It's a three-part sport: obedience, tracking, and protection work. For more on Schutzhund trials in the United States, see http://www.germanshepherddog.com/schutzhund/trial.htm.

For the section on singleton litters, I depended on Patricia McConnell, one of my favorite animal behaviorists, who wrote about her experience with a single pup (also named Solo) in *For the Love of a Dog: Understanding Emotion in You and Your Best Friend* (New York: Ballantine Books, 2006). As she put it in her notes, her experience is simply anecdotal, with little scientific research to back it up. "That doesn't mean it has no value, but until someone does carefully controlled research on the topic, we need to be cautious about drawing conclusions." She placed her Solo with a single woman, and he was "the light of her life."

I also found Karen London's Dog Behavior Blog, on singleton puppies, helpful: http://www.dogbehaviorblog.com/2008/08/singleton-puppi.html, last modified August 2, 2008.

On Solo's ability to vocalize: I don't believe he is closely related to the wild dogs of Africa; nonetheless, that species has been classified as among the most social of all canids, a category that includes wolves, foxes, coyotes, jackals, and dogs. The wild dogs of Africa have one of the most complex vocal repertoires of all the canids.

2: Death and the Dog

Interviews, correspondence, personal communications, and background information for this chapter include archaeologist Haleh Brooks; Dr. Edward David, former deputy chief medical examiner for the state of Maine; Brad Dennis, search director for KlaasKids Foundation; Lisa Higgins, cadaver-dog trainer and handler from Louisiana, and a civilian contractor with the FBI on the victim recovery team; Nancy Hook; Paul Martin, anthropology graduate student at the University of Mississippi and cadaver-dog program coordinator for the Western Carolina University Forensic Anthropology Program; Lisa Mayhew, child death investigator/trainer for the North Carolina office of the chief medical examiner; Andrew "Andy" Rebmann, retired Connecticut state trooper and founder of K-9 Specialty Search Associates; and research associate professor Marcella Sorg, Department of Anthropology, University of Maine.

The epigraph for this chapter comes from *The Others: How Animals Made Us Human,* by Paul Shepard (Washington, D.C.: Island Press, 1995). Shepard, a deep ecologist, wrote of the dog, "It is a borderline animal in so many ways that its marginality has mythic proportions, especially in connection with the geography of chaos" (62).

The book most central to this chapter—indeed, to this entire book—is *Cadaver Dog Handbook: Forensic Training and Tactics for the Recovery of Human Remains,* by Andrew Rebmann, Edward David, and Marcella Sorg (Boca Raton, FL: CRC Press, 2000). It is considered the central text for cadaver-dog handlers and trainers. The other book that Nancy Hook mentioned in our first conversation about scenting dogs was William G. Syrotuck's 102-page book, *Scent and the Scenting Dog,* first published in 1972 (Rome, NY: Arner Publications). It has been reprinted numerous times and remains one of the classics on nose work.

The section on the history of dogs and death opens with the discovery of the canid skulls (there is some archaeological back-and-forth about whether those skulls represent "the dog"): "Palaeolithic Dog Skulls at the Gravettian Predmostí Site, the Czech Republic," by Mietje Germonpré, Martina Láznicková-Galetová, and Mikhail V. Sablin, *Journal of Archaeological Science* 39, no. 1, 2012: 184–202.

Whether Egypt's god Anubis is a desert dog or a jackal remains unknown, but as Stephanie Cass notes, the figure is "definitely canid and most likely a jackal or a wild dog—or a hybrid of both—but, as in the case of Seth (the god of chaos), with alterations that deliberately smudge the lines of reality." ("Anubis," *Encyclopedia Mythica Online,* http://www.pantheon.org/articles/a/anubis.html, last modified January 16, 2004.)

The quote on *canes sepulchrales* comes from *Introduction to the Study of Mortuary Customs Among the North American Indians,* by Dr. H. C. Yarrow (Washington, D.C.: Smithsonian Institution, Bureau of Ethnology, 1880: 10). It's one of the few sources available on these dogs, but it comports with what is known about the habits of dogs.

The Zoroastrian *sagdid* is well known and documented, with numerous sources and research devoted to it, including *The Religious Ceremonies and Customs of the Parsees,* by Jivanji Jamshedji Modi (Mazagon, Bombay, British India Press, 1922), http://www.avesta.org/ritual/rcc.htm, accessed November 2011.

I am especially indebted to Haleh Brooks, an archaeologist now living in Norway, who took a snapshot in the museum of Tehran of a small bronze dog from ancient Persia that looks remarkably like a German shepherd. Her blog post of February 2012, http://halehsworldofarchaeology.blogspot.com/2012/02/dogs-in-ancient-iran.html, was enormously helpful, as has been her correspondence.

British scholar Mary Boyce, who died in 2006, is still considered the world's foremost authority on Zoroastrianism and wrote extensively on the role of dogs in that religion. One of her easily accessible pieces on the web is "Fauna and Flora: Dog in Zoroastrianism," an excerpt from *Encyclopaedia Iranica,* http://www.cais-soas.com/CAIS/Animals/dog_zoroastrian.htm, accessed December 13, 2011.

The section on the role of dogs in early Western civilization and monotheistic religions has a number of sources, but the article by historian Sophia Menache at the University of Haifa, "Dogs: God's Worst Enemies?" *Society & Animals Journal of Human-Animal Studies* 5, no. 1, 1997: 23–44, was most helpful, as was the article "Guardians of the Corpse Ways," by amateur historian Robert N. Trubshaw, *Mercian Mysteries* no. 20, August 1994, http://www.indigogroup.co.uk/edge/bdogfl.htm, last modified November 2008.

Scamp, "The Grim Reaper Dog," was featured in newspapers and television stories across the United States and Great Britain, including *Extraordinary Animals* on

Animal Planet, which originally aired May 31, 2011, http://www.thepetnetwork .tv/videos/extraordinary-animals-the-grim-reaper-dog/, accessed November 2011.

Numbers for "black dog syndrome" are difficult to come by—many shelter managers say it's a problem, but a few people call it an urban legend. On black dogs at shelters, see, for example, "Black Dogs Face a Hard Choice at Shelter," by Deb Hipp, *The Bark* 35, March/April 2006, and "Black Dog Bias?" by Craig Nakano, *Los Angeles Times,* December 6, 2008, http://www.latimes.com/features/la-hm -black6-2008dec06,0,6461430.story, accessed November 2011.

Tales of discovering bodies while walking one's dog abound in the mainstream media. Lauren Kornberg was interviewed for *The Madeleine Brand Show:* "Hollywood Severed Head: Dog Walker Recounts Grim Discovery," 89.3 KPCC, Southern California Public Radio (January 19, 2012), http://www.scpr.org/programs/ madeleine-brand/2012/01/19/22180/lauren-kornberg-hollywood-human-head-sev ered, accessed March 2012; Fish, the dog, was featured in "Pet Dog's Discovery of Decaying Hand Leads Police to Body in Mission," by Naxiely Lopez, *The Monitor,* http://www.themonitor.com/articles/mission-53721-human-carrying .html#_jmp0_, last modified August 12, 2011.

There is a limited but increasing academic literature on canine, coyote, and bear predation on human remains. Here is a brief sampling of the articles and textbook chapters devoted to the subject: "Canid Modification of Human Remains: Implications for Time-Since-Death Estimations," by P. Wiley and L. M. Snyder, *Journal of Forensic Sciences* 34, no. 4, 1989: 894–901; "Canid Scavenging/Disarticulation Sequence of Human Remains in the Pacific Northwest," by William D. Haglund, Donald T. Reay, and Daris R. Swindler, *Journal of Forensic Sciences* 34, no. 3, 1989: 587–606; "Dogs and Coyotes: Postmortem Involvement with Human Remains," by William D. Haglund, in *Forensic Taphonomy: The Postmortem Fate of Human Remains,* edited by William D. Haglund and Marcella H. Sorg (Boca Raton, FL: CRC Press, 1997): 367–381; *Skeletal Manifestations of Bear Scavenging,* by E. Ann Carson, Vincent H. Stefan, and Joseph F. Powell, *Journal of Forensic Sciences* 45, no. 3, 2000: 515–526; "Taphonomy of Child-Sized Remains: A Study of Scattering and Scavenging in Virginia, USA," by Robert Morton and Wayne Lord, *Journal of Forensic Sciences* 51, no. 3, 2006: 475–479.

Archaeologists may have underestimated dogs' roles at digs, misreading dog activity as human activity: "Analogic Reasoning, Ethnoarchaeology, and the Impact of Canines on the Archaeological Record," by Lawrence A. Kuznar and Robert Jeske,

Archeological Papers of the American Anthropological Association 16, 2006: 37–46; and "Identifying the Involvement of Multiple Carnivore Taxa with Archaeological Bone Assemblages," by Marie M. Selvaggio and J. Wilder, *Journal of Archaeological Science* 28, no. 5, 2001: 465–470. Even something as simple as the joy of digging pits in the ground may have been misinterpreted: "Canine Digging Behavior and Archaeological Implications," by Robert J. Jeske and Lawrence A. Kuznar, *Journal of Field Archaeology* 28, nos. 3–4, 2001: 383–394.

Some of the saddest cases are those of lonely people whose remains are found in houses or apartments with dogs who have inevitably eaten their deceased owners: "Case Report: Canine Scavenging of Human Remains in an Indoor Setting," by Dawnie Wolfe Steadman and Heather Worne, *Forensic Science International* 173, 2007: 78–82.

3: Nose Knowledge

This chapter includes material from interviews with K9 legal expert Terry Fleck; director emeritus of the International Forensic Research Institute at Florida International University Kenneth Furton; cognitive psychologist William S. "Deak" Helton at the University of Canterbury in New Zealand; Marcia Koenig, of K9 Specialty Search Associates; Nicholas "Nick" Montanarelli, who worked for numerous government agencies during his career, including the U.S. Army Land Warfare Laboratory; associate professor of animal behavior and sensory physiology and medicine at Auburn University Laurence "Larry" Myers; retired Maine game warden and owner of Maine K-9 Services Deborah Palman; Andy Rebmann; retired neurosurgeon George Stevenson, who took up the grizzly brain after his retirement; and Roger Titus, vice president of the National Police Bloodhound Association.

This chapter benefited from my observation of Roger Titus training bloodhound handlers. I have watched dozens of training sessions with patrol dogs from the Durham Police Department K9 unit and the Durham Sheriff K9 unit, among many other law enforcement agencies, mostly in North Carolina.

Also critical to this chapter were Marcia Koenig's unpublished papers; Andy Rebmann's news clip book with cases involving him, his dogs, and law enforcement handlers he trained going back to the early 1970s; and Nick Montanarelli's snapshots of some of the early military training and research with dogs during the 1960s and '70s.

Books critical to this chapter include *Cadaver Dog Handbook* and Deak Helton's edited collection, *Canine Ergonomics: The Science of Working Dogs* (Boca Raton, FL: CRC Press, 2009). Leon Whitney—veterinarian, breeder of bloodhounds, and enthusiastic proponent of stuffing purebred dogs and exhibiting them at the Yale Peabody Museum—authored the classic *Bloodhounds and How to Train Them* (New York: Orange Judd Publishing Co., 1947). The book by recently deceased K9 trainer Tracy Bowling, *Police K9 Tracking: A Guide for Training & Deploying the Police Tracking Dog* (K9 Publishing, 2010), was also helpful. For understanding more about humans' sense of smell, Avery Gilbert's *What the Nose Knows: The Science of Scent in Everyday Life* (New York: Crown Publishing, 2008) was a guide. Although I don't agree with a few of John Bradshaw's conclusions about dog cognition, his book *Dog Sense: How the New Science of Dog Behavior Can Make You a Better Friend to Your Pet* (New York, Basic Books, 2011) is chock-full of wonderful history and science.

The article on the shrew breakthrough was "Fossil Evidence on Origin of the Mammalian Brain," by Timothy B. Rowe, Thomas E. Macrini, and Zhe-Xi Luo, *Science* 332 (2011): 955–957. Animal Planet's overtouting of the bloodhound's nose can be found on its website under "Top 10 Animal Skills," http://animal.discovery.com/tv/a-list/creature-countdowns/skills/skills-09.html, accessed November 2011.

Frank C. Craighead Jr. and his identical twin, John, are credited with helping save the grizzly bear from extinction with their twelve-year study of radio-collared grizzlies around Yellowstone. As the *New York Times* noted in its obituary of Frank Craighead, he and his brother had to go the extra (vertical) mile in their research: "They and the students who worked with them did pull-ups and other calisthenics to build strength in case they had to climb a tree to elude a grizzly." The paper that shows up with a certain promiscuity is "Grizzly Bear Ranges and Movement as Determined by Radiotracking," by Frank C. Craighead Jr., Third International Conference on Bears, Paper 10, 1976: 97–109.

As noted, even Nobel Prize winner Linda B. Buck has a hard time getting specific about what humans can detect. Her quote comes from "Olfactory Receptors and Odor Coding in Mammals," *Nutrition Reviews* 62, no. 11, 2004: 184–188. There's extensive literature on physicians' evaluation of "effluvia": see, for instance, "Advances in Electronic-Nose Technologies Developed for Biomedical Applications," by Alphus D. Wilson and Manuela Baietto, *Sensors* 11, 2011: 1,105–1,176. On Larry Sunshine, odor mitigation expert, see "On PATH Trains, Noses Wrinkle at a Moldy Mystery," *New York Times,* June 7, 2011: A22. The quote on patchouli comes from *Perfumes: The Guide,* by Luca Turin and Tania Sanchez (New York: Viking Adult, 2008), 283–284.

The idea that trained noses are better than untrained noses isn't an epiphany to anyone in the sniffer-dog world, nor is it a surprise to enologists or perfumers. However, the chocolate-tracking study—"Mechanisms of Scent-tracking in Humans," by Jess Porter, Brent Craven, Rehan M. Khan, Shao-Ju Chang, Irene Kang, Benjamin Judkewitz, Jason Volpe, Gary Settles, and Noam Sobel, *Nature Neuroscience* 10, no. 1, 2007: 27–29—probably got extensive mainstream media attention because it was also an article about chocolate. The water-taste study was designed to test the popular belief that the visually impaired have a better sense of smell than others. The study, "Smell and Taste Function in the Visually Impaired," by Richard S. Smith, Richard L. Doty, Gary K. Burlingame, and Donald A. McKeown, *Perception & Psychophysics* 54, no. 5, 1993: 649–655, showed that the best performers were not the visually impaired but the trained and sighted employees of the Philadelphia Water Department who served on water-quality panels.

The ever-growing listing of scent-discrimination tasks for dogs comes from both studies and media reports. The majority of these sniffer tasks don't have peer-reviewed, scientifically controlled studies associated with them.

The best history of police dogs is *Police Dogs in North America,* by Samuel G. Chapman (Springfield, IL: Charles C. Thomas, 1990). Marcia Koenig's unpublished paper, "History of Search Dogs in the United States" (1989, updated in 2009), was helpful as well.

The scent-discrimination studies that look at how dogs are able to differentiate between identical twins will probably continue to develop and be challenged. The first person to test the hypothesis and find dogs lacking was Hans Kalmus, in "The Discrimination by the Nose of the Dog of Individual Human Odours and in Particular the Odours of Twins," *British Journal of Animal Behavior* 3, no. 1, 1955: 25–31. Kalmus did note that the dogs used for the study varied greatly "in intelligence, perseverance and the degree to which they had been trained." Czech scientists Ludvik Pinc, Luděk Bartoš, Alice Reslová, and Radim Kotrba came up with a very different result in "Dogs Discriminate Identical Twins," *PloS ONE* 6, no. 6, 2011: 1–4, http://www.plosone.org/article/info:doi/10.1371/journal.pone.0020704, accessed November 2011.

NOTES

The quote about dogs' receptor cells comes from Stanley Coren's book, *How Dogs Think: Understanding the Canine Mind* (New York: Free Press, 2004), 55.

The report that Lester Shubin's quotes are included in is "25 Years of Criminal Justice Research," National Institute of Justice, Report No. 151287, December 1994.

The section on bloodhounds contains material from numerous scholarly and trainer sources, including historian John Campbell's fascinating work "The Seminoles, the 'Bloodhound War,' and Abolitionism, 1796–1865," *The Journal of Southern History* 62, no. 2, 2006: 259–302; and Christina Chia's "Rethinking the Slave Hunt: On the Trail of Humans and Dogs in American Slavery," American Studies Association, 2008. Deborah Palman's quote about bloodhounds comes from her web piece, *Negative Searching,* http://emainehosting.com/mesard/Articles/Negative%20Searching.htm.

You can see Walt Disney's 1930 cartoon *The Chain Gang* on YouTube: http://www.youtube.com/watch?v=803r5DzEqb4.

4: Birth of the Body Dog

Interviews and correspondence critical to this chapter include those with Edward David; Joan Johnston, wife of military researcher William H. Johnston; Marcia Koenig; Nick Montanarelli; Deborah Palman; James J. Polonis, former Southwest Research Institute project manager; Andy Rebmann; Marcella Sorg; and retired New York State Trooper Ralph D. "Jim" Suffolk Jr. and his wife, Sally Suffolk.

Books used for this chapter include *Cadaver Dog Handbook; The War Animals,* by Robert E. Lubow (New York: Doubleday, 1977); *A Killer Named Hatch: Massacre on Potato Hill: A True Story,* by Thomas Blanchfield (AuthorHouse, 2009); *Manhunters!: Hounds of the Big T,* by William D. Tolhurst as told to Lena F. Reed (Puyallup, WA: Hound Dog Press, 1984); and *K-9 Cops: Stories from America's K-9 Police Units,* by Richard Rosenthal (New York: Pocket Books, 1997).

A number of declassified military research reports were also important to this chapter, among them "Body Recovery Dog," by Woodrow L. Quinn Jr. and Nicholas Montanarelli, U.S. Army Land Warfare Laboratory, Aberdeen Proving Ground, MD, Technical Report No. IRL-03B73, May 1973; "Mine Detecting Canines," by R. V. Nolan and D. L. Gravitte, U.S. Army Mobility Equipment Research and Development Command, Fort Belvoir, VA, Technical Report No. 2217, September

1977; and Southwest Research Institute's "Interim Technical Report on Olfactory Acuity in Selected Animals Conducted During the Period June 1972–September 1974," Report No. AD-787 495, September 1974.

Material for the section on Jim Suffolk, in addition to interviews and documents from his own archives, include the legal case *The People of the State of New York* v. *Ralph Centolella and John Pella,* County Court of New York, Oneida County 61 Misc. 2d 726; 305 N.Y.S. 2d 460, 1969.

The *Hartford Courant* was one of the major newspapers to cover Andy Rebmann and his K9s, but the *Norwich Bulletin*; the *Morning Union*; the *Day* of New London, Connecticut; the *Providence Journal Bulletin*; the *Boston Globe*; and several magazines also reported on his work. In addition to interviewing Andy, I used many of those newspaper, magazine, and web sources, along with Rosenthal's book, *K-9 Cops,* to create the portraits of the more well-known cases Andy was involved in, including Robin Oppel's disappearance in Monroe, Connecticut, and the murder and dismemberment of Helle Crafts. Material for the section on the L'Ambiance Plaza collapse included a number of *Courant* articles. The *Courant* sent twenty reporters and photographers (including me) to cover the collapse and its aftermath. *Courant* reporter Lynne Tuohy especially covered the aspect of the search dogs deployed in this disaster.

Thomas B. Slick Jr.'s patent, "Apparatus for Erecting a Building, US Patent 2,715,013," was filed July 6, 1948, and issued August 19, 1955, http://www.google .com/patents?vid=2715013, accessed December 2011.

The Korean Air crash in Guam in 1997 that Marcia Koenig worked with her dog Coyote killed 228 of the 254 people aboard. Pilot fatigue, lack of crew training, and an outdated flight map contributed to the disaster.

5: The Shell Game

Interviews, correspondence, and training observations important to this chapter include those with Joan Andreasen-Webb; Mike Baker; Kevin George, president of Sentry Dog Services, Ltd.; Deak Helton; Lisa Higgins; working-dog trainer and handler Kathy Holbert of Philippi, West Virginia; Nancy Hook; Marcia Koenig; and University of California, Davis, researcher Dr. Lisa Lit.

Books used for this chapter include *Cadaver Dog Handbook* and *Canine Ergonomics.* Kevin George's paper and program, "A Box Is a Box Is a Box," is copyrighted.

6: Distillations

Interviews and correspondence and observations critical to this chapter include those with Dr. Mary E. Cablk, associate research professor at the Desert Research Institute; Ken Furton; Lisa Higgins; Nancy Hook; Paul Martin; Lisa Mayhew; Andy Rebmann; Dr. John C. Sagebiel, University of Nevada, Reno; and forensic anthropologist Arpad A. Vass. I also benefited from visits and training at FOREST, Western North Carolina's outdoor human decomposition research facility.

The epigraph is from Cincinnati *Clinic* magazine (September 4, 1875), "Odor Mortis; or, the Smell of Death." Dr. A. B. Isham read that paper in front of the Cincinnati Academy of Medicine on August 30, 1875.

Cadaver Dog Handbook remains the most practical, thorough, and complete textbook on these issues. Academic articles important to this chapter include "Beyond the Grave—Understanding Human Decomposition," by Arpad A. Vass, *Microbiology Today* 28, November 2001: 190–192; "Odor Analysis of Decomposing Buried Human Remains," by Arpad A. Vass, Rob R. Smith, Cyril V. Thompson, Michael N. Burnett, Nishan Dulgerian, and Brian A. Eckenrode, *Journal of Forensic Sciences* 53, no. 2, 2008: 384–391; "Characterization of the Volatile Organic Compounds Present in the Headspace of Decomposing Human Remains," by Erin M. Hoffman, Allison M. Curran, Nishan Dulgerian, Rex A. Stockham, and Brian A. Eckenrode, *Forensic Science International* 186, 2009: 6–13; and "Characterization of the Volatile Organic Compounds Present in the Headspace of Decomposing Animal Remains, and Compared with Human Remains," by Mary E. Cablk, Erin E. Szelagowski, and John C. Sagebiel, *Forensic Science International* 220, 2012: 118–125.

There are only six or seven scientific studies on cadaver dogs, along with two controlled studies. The one I found most evocative was that of Lars Oesterhelweg and his fellow researchers, "Cadaver Dogs—A Study on Detection of Contaminated Carpet Squares," by L. Oesterhelweg, S. Kröber, K. Rottmann, J. Willhöft, C. Braun, N. Thies, K. Püschel, J. Silkenath, and A. Gehl, *Forensic Science International* 174, 2008: 35–39. Mary Cablk and John Sagebiel authored a well-controlled study, "Field Capability of Dogs to Locate Individual Human Teeth," *Journal of Forensic Sciences* 56, no. 4, 2011: 1,018–1,024. Earlier studies included those of Debra Komar, "The Use of Cadaver Dogs in Locating Scattered, Scavenged Human Remains: Preliminary Field Test Results," *Journal of Forensic Sciences* 44, no. 2,

1999: 405–408; and "Cadaver Dog and Handler Team Capabilities in the Recovery of Buried Human Remains in the Southeastern United States," by Alanna E. Lasseter, Keith P. Jacobi, Ricky Farley, and Lee Hensel, *Journal of Forensic Sciences* 48, no. 3, 2003: 617–621.

Although people kept asking me if I planned to see the science-museum hit *Bodies: The Exhibition,* I found the idea disturbing. The ethical issues were nicely summarized by Renée Marlin-Bennett, Marieke Wilson, and Jason Walton in "Commodified Cadavers and the Political Economy of the Spectacle," *International Political Sociology* 4, 2010: 159–177: "Notwithstanding the enthusiastic response of the spectators (or because of it), these claims of educational value seem to put a gloss of respectability and virtue on these productions while silencing the unsettling questions: What is the provenance of the bodies? Is it ethical to use human mortal remains for commercial purposes? Why do the rules governing the international transportation of cadavers not apply to these plastinated dead? . . . Indeed, the spectacle hides a silence in international and domestic laws governing an extreme form of human trafficking—trafficking in dead humans." (160)

7: A Spare Rib

Interviews and correspondence for this chapter include those with Joan Andreasen-Webb; Roy and Suzie Ferguson of Tennessee Special Response Team-A; Suzi Goodhope of Southeast K-9 Search and Rescue and KlaasKids search team; Nancy Hook; Marcia Koenig; and Andy Rebmann. Team members of the Triad Bloodhound Team helped enormously with early training of Solo, especially Ken Young and Darlene Griffin. I also benefited from observing at seminars, including Dogmeet 2011 at the Kitsip County Fairgrounds in Bremerton, Washington.

I used a training record model from Andy's *Cadaver Dog Handbook* for Solo, which included time, temperature, humidity, wind direction, what material we used, where we hid the material, and how both Solo and I did. Those records helped with time lines and training descriptions for this book.

In retrospect, the decision David and I made to neuter Solo probably wasn't the answer to Solo's problems; we are now learning about a multitude of health reasons to delay or never neuter male and female dogs, despite the pressures from national and local spay-and-neuter programs. It's possible that, because of early neutering, Solo grew significantly taller and didn't develop some of the muscle he should have. One of the clearest explanations of the risks and benefits of neutering, especially for active dogs, comes from veterinarian Chris Zink: Her PDF article, "Spay Neuter Considerations, 2013," can be downloaded from http://www.caninesports.com/useful-info.html.

Bruce Lee's quote comes from the film *Enter the Dragon,* directed by Robert Clouse (Hollywood, CA: Warner Brothers, 1973).

8: Comfort Me with Bite Work

Interviews and correspondence for this chapter include those with Mike Baker; K9 legal expert Terry Fleck; Nancy Hook; Broward Sheriff's Deputy Dave Lopez; Lisa Mayhew; Dr. Charlie Mesloh of Florida Gulf Coast University; Indianapolis Metropolitan Police K9 training supervisor Craig Patton; Andy Rebmann; Broward Sheriff's Deputy Peter Sepot; and Steve Sprouse, patrol-dog trainer for the Broward Sheriff's Office in Southeast Florida. For the past seven years, I have been fortunate to be able to observe suspect apprehension training under a number of law enforcement K9 trainers in both North Carolina and Florida, at standard weekly trainings, and at seminars.

The book *Police and Military Dogs: Criminal Detection, Forensic Evidence, and Judicial Admissibility,* by John Ensminger (Boca Raton, FL: CRC Press, 2011), was helpful, along with numerous *Police K-9* magazine and *K-9 Cop* magazine feature stories and columns. Other articles for this chapter include "Police Service Dogs in the Use-of-Force Continuum," by Jonathan K. Dorriety, *Criminal Justice Policy Review* 16, no. 1, 2005: 88–98; "Barks or Bites? The Impact of Training on Police Canine Force Outcomes," by Charlie Mesloh, *Police Practice and Research* 7, no. 4, 2006: 323–335; "Practical, Ethical and Political Aspects of Engaging 'Man's Best Friend' in the War on Crime," by B. Grant Stitt, *Criminal Justice Policy Review* 5, no. 1, 1991: 53–65; "Changes in Officer Use of Force over Time: A Descriptive Analysis of a National Survey," by Bruce Taylor, Geoffrey Alpert, Bruce Kubu, Daniel Woods, and Roger G. Dunham, *Policing: An International Journal of Police Strategies & Management* 34, no. 2, 2011: 211–232.

The courts have consistently found that patrol dogs used for apprehension do not constitute the use of deadly force. *Robinette* v. *Barnes,* 854 F.2d 909 (6th Cir. 1988), is one of the earliest opinions. During a search of a car dealership because an alarm had gone off, a handler sent his dog into a darkened area after giving the K9 warning. The suspect was lying under a car, and the dog bit him in the neck, killing him. This is the only major legal case at this point involving a dog fatality. There are, however, numerous lawsuits contending "excessive force" in using dogs for suspect apprehension. Sometimes entire departments come under justified scrutiny, such as Prince George's County police K9 unit in the 1990s. The *Washington Post,* especially reporter Ruben Castaneda, did numerous stories on the Prince George K9 unit, including, for example, "FBI Probing Canine Unit; Lawsuits Recount Attacks by Pr. George's Police Dogs" (April 4, 1999), A1. By 2007, the K9 unit, which had operated under a consent decree, improved to the extent that a federal judge ruled the unit no longer needed oversight.

The section on Steve Sprouse being shot depended on a number of sources, including Steve himself, but also newspaper articles at the time of the shooting, as well as a court case filed by the arrested and charged defendant. That case was dismissed.

9: Into the Swamp

Interviews and correspondence include those with Mike Baker; Deborah Palman; Milo Pyne, NatureServe's senior regional ecologist for the southeastern United States; and North Carolina State University botany doctoral student Wade Wall. Ellerbee Creek Watershed Association's naturalist tours of the Durham area also helped me re-create the flora of this search area. Though I did not use names in this account, no facts were changed.

Two books were helpful, especially *Wildflowers and Plant Communities of the Southern Appalachian Mountains and Piedmont,* by Timothy P. Spira (Chapel Hill: University of North Carolina Press, 2011); and *Field Guide to the Piedmont,* by Michael A. Godfrey (Chapel Hill: University of North Carolina Press, 1997).

Deborah Palman's account of distance alerts comes partly from a lecture she gave at the National Search Dog Alliance February 2011 conference, in Eatonton, Georgia, and from her article, " 'Distant Alerts'—Long Distance Scent Transport in Searches for Missing Persons," *USPCA Canine Courier* 22, no. 1, March 2011: 47–51. It can be downloaded from emainehosting.com/mesard/pdf_documents/Distant%20 Alerts.pdf.

10: Cleverness and Credulity

Interviews and correspondence for this chapter include those with Mike Baker, Terry Fleck, *Texas Tribune* managing editor Brandi Grissom, Lisa Lit, Larry Myers,

Andy Rebmann, and Roger Titus, as well as numerous personal communications with handlers and trainers, and my own training experience and observation.

Books important to this chapter include Harry G. Frankfurt's punchy *On Bullshit* (Princeton, NJ: Princeton University Press, 2005), originally published as an article in *Raritan Quarterly Review* 6, no. 2, Fall 1986; Oskar Pfungst's *Clever Hans (The Horse of Mr. Von Osten): A Contribution to Experimental Animal and Human Psychology* (New York: Henry Holt and Company, 1911), available at http://www.gutenberg.org/files/33936/33936-h/33936-h.htm; Bowling's *Police K9 Tracking;* Ensminger's *Police and Military Dogs,* as well as Ensminger's blog: http://doglaw reporter.blogspot.com/. The two *New York Times* stories about bed bugs were "A New Breed of Guard Dog Attacks Bedbugs," by Penelope Green (March 10, 2010), Home and Garden section; and "Doubts Rise on Bedbug-Sniffing Dogs," by Cara Buckley (November 11, 2010), New York Region News section.

The piece on dog alerts, "Handler Beliefs Affect Scent Detection Dog Outcomes," by Lisa Lit, Julie B. Schweitzer, and Anita M. Oberbauer, was published in *Animal Cognition* 14, no. 3, 2011: 387–394; Lit has also written on search-dog and handler performance: "Effects of Training Paradigms on Search Dog Performance," Lisa Lit and Cynthia A. Crawford, *Applied Animal Behaviour Science* 98, 2006: 277–292.

The sections on John Preston and Keith Pikett depended on lengthy investigative journalist reports, especially from *Florida Today* and the *Texas Monthly.* The *New York Times* also covered the Pikett trials. The Innocence Project of Texas report, "Dog Scent Lineups: A Junk Science Injustice" (September 21, 2009), can be accessed at http://www.ipoftexas.org/documents; the CNN videos of Keith Pikett with his dogs can be accessed at http://www.cnn.com/2009/CRIME/10/05/texas .sniffer.dogs.controversy/#cnnSTCVideo.

The section on Sandra M. Anderson utilized numerous news sources. The fulsome language of the *Archaeology* magazine article entitled "Hounding the Dead: A Remarkable Michigan Mutt Sniffs out Ancient Human Remains," by Brenda Smiley, *Archaeology* 53, no. 5, 2000, was followed several years later by an online feature written by an intern: "Canine Case Closed?" about Sandra Anderson's conviction, http://archive.archaeology.org/online/features/dogs/index.html, accessed November 2011. Two research papers on the web were especially helpful: "No, Your Friend Cannot Do Magic: *United States* v. *Sandra Marie Anderson and Cadaver Dogs on Trial,*" by Liz Burne, http://www.searchdogsne.org/reference.html, accessed December 2011; and "Fraudulent Use of Canines in Police Work," by Daniel A.

Smith, Lincoln Park Police Department, http://ebookbrowse.com/fraudulent-use
-of-canines-in-police-work-pdf-d18075497, accessed December 2011.

11: All the World's a Scenario

Interviews and correspondence for this chapter include those with Brad Dennis, search director for KlaasKids Foundation; Roy and Suzie Ferguson; Lisa Higgins and her granddaughter Haylee; Nancy Hook; Benjamen Ortiz; Andy Rebmann; Arpad Vass; and Roane County Sheriff Detective Art Wolff, founder of Tennessee Special Response Team-A. Seminars where scenarios were used included a February 2011 National Search Dog Alliance seminar in Eatonton, Georgia; a September 2011 Washington Explorer Search and Rescue (WESAR) DogMeet in Bremerton, Washington; an October 2011 Network of Canine Detection Services Working Dog Seminar in Holly Springs, Mississippi; a November 2011 Cadaver Dog Workshop at Western Carolina University; and an August 2011 training with Tennessee Special Response Team-A in Sevierville, Tennessee.

The scenarios that Art Wolff and his team develop are sophisticated and invaluable, and I am appreciative of their generosity in letting me write about them.

The Winthrop Point isn't a widely advertised phenomenon; indeed, it doesn't exist on the web.

The scenario that Brad Dennis used to help educate handlers was based on the rape and murder of Chelsea King in San Diego in 2010: see http://www.cnn.com/2010/CRIME/04/17/chelsea.king.gardner.plea/index.html and "Creation of a Monster, John Gardner," by Don Bauder, *San Diego Reader,* http://www.sandiegoreader.com/news/2012/jun/27/citylights1-creation-monster-john-gardner/, accessed June 2012. The Washington State study cited by Brad Dennis is *Case Management for Missing Children Homicide Investigation, Report II,* by Katherine M. Brown, Robert D. Keppel, Joseph G. Weis, and Marvin Skeen (published by cooperative agreement between Rob McKenna, attorney general of Washington, and the U.S. Department of Justice Office of Juvenile Justice and Delinquency Prevention, 2006), http://www.atg.wa.gov/ChildAbductionResearch.aspx#_jmp0_, accessed December 2011.

12: The Grief of Others

This chapter is based on my own search experience, my academic and professional knowledge of the media, personal conversations with numerous other cadaver-dog handlers and trainers, and correspondence and interviews with Brad Dennis, Nancy Hook, Lisa Mayhew, Andy Rebmann, and Arpad Vass, as well as visits and training at Western North Carolina's FOREST facility.

Books and articles, in addition to the always helpful *Cadaver Dog Handbook,* include *Analysis of Lost Person Behavior,* by William Syrotuck and Jean Anne Syrotuck (Mechanicsburg, PA: Barkleigh Productions Inc., 2000); "The Lost Alzheimer's and Related Disorders Subject: New Research and Perspectives," by Robert J. Koester, *Response 98 NASAR Proceedings,* 1998: 165–181; "Behavioral Profile of Possible Alzheimer's Disease Subjects in Search and Rescue Incidents in Virginia," by Robert J. Koester and David E. Stooksbury, *Wilderness and Environmental Medicine* 6, 1995: 34–43; and "The Search for Human Remains in the Search and Rescue Environment," by Mark Gleason, Search and Rescue Tracking Institute, Virginia, February 2008.

The National Crime Information Center's list of "endangered" missing people for 2012 can be found at http://www.fbi.gov/about-us/cjis/ncic/ncic-missing-person -and-unidentified-person-statistics-for-2012. Crime rates in North Carolina can be tracked at http://www.disastercenter.com/crime/nccrimn.htm.

The section on New Bedford, Massachusetts, was greatly aided by Carlton Smith's book *The Killing Season: The Unsolved Case of New England's Deadliest Serial Killer* (New York: Onyx, 1994) and Rosenthal's *K-9 Cops.* Archived newspaper clippings from the *Hartford Courant* and other local papers from the period were helpful, as well as Denise Dowling's article "Speaking for the Dead," *Rhode Island Monthly,* May 2009, http://www.rimonthly.com/Rhode-Island-Monthly/May-2009/ Speaking-for-the-Dead/#_jmp0_, accessed December 2011; and Curt Brown's series for the North Kingstown, Rhode Island, newspaper, the *Standard-Times,* "Dark Days Revisited: Bristol County Highway Killings, 20 Years Later," July 7, 2008, http://www.southcoasttoday.com/apps/pbcs.dll/article?AID=/20080707/ NEWS/807070337, accessed December 2011.

13: All the Soldiers Gone

Interviews and correspondence for this chapter included Kathy Holbert; University of Rhode Island associate professor and documentary filmmaker Mary Healy Jamiel; Andy Rebmann; Greg Sanson, personnel recovery advisor to the U.S. military in Iraq; freelance writer and journalist Michael Sledge; Rhode Island State Trooper Matt Zarrella; and a number of personal communications with handlers, trainers, and military working-dog personnel.

The epigraph by the World War I hero Siegfried Sassoon reflects his increasing disillusionment with the war. The Oxford University Press biography for its collection of his works notes: "[D]uring his recovery period, discouraged by the politics of war at home and the deaths of numerous friends at the front, he made contact with the group of pacifists led by Bertrand Russell and Lady Ottoline Morrell. He returned to France in January 1917, was wounded by a sniper during a raid near Fontaine-les-Croisilles in April, and was sent back to England. Encouraged by Russell and the journalists John Middleton Murry and H. W. Massingham, he wrote his soldier's statement, dated 15th June 1917, calling for a negotiated peace, and acted to resign his commission. Robert Graves intervened, fearing that his friend would be court-martialed. His commanding officers were sympathetic, and sent Siegfried to Craiglockhart War Hospital near Edinburgh to be treated for neurasthenia."

Books that were important to this chapter include Michael Sledge's *Soldier Dead: How We Recover, Identify, Bury, and Honor Our Military Fallen* (New York: Columbia University Press, 2005) and Drew Gilpin Faust's *This Republic of Suffering: Death and the American Civil War* (New York: Knopf, 2008).

Ellwood Hendrick's "Merciful Dogs of War" ran in the February 1917 issue of *Red Cross Magazine*. Although I could not access the original magazine article, it is cited and reproduced in several newspapers and magazines from the early months of 1917, including the *Oakland Tribune,* February 20, 1917: 10, under the heading "Tale of the Red Cross Dog." Theo. F. Jager's eighty-three-page book, *Scout, Red Cross and Army Dogs: A Historical Sketch of Dogs in the Great War and a Training Guide for the Rank and File of the United States Army* (New York: Arrow Printing Co., 1917) can be accessed via googleplay: https://play.google.com/store/books/details?id=hCUwJ8dyzcYC.

Historian J. David Hacker's work has been seminal—and is thought to be considerably more accurate—on revised counts of deaths in the Civil War. *Civil War History* journal editors, in awarding his piece "A Census-Based Count of the Civil War Dead," *Civil War History* 57, no. 4, 2011: 307–348, the best of 2011, wrote: "These results have far-reaching consequences, encouraging historians to rethink assumptions not only about the war's human cost, but the ways in which we try to measure and comprehend the size of that cost."

Ernie Pyle's draft column can be read at the Indiana University School of Journalism's website dedicated to Ernie Pyle: http://journalism.indiana.edu/resources/erniepyle/. Pyle is one of my heroes, and I still get emotional reading his work. In his photographs, he resembled my grandfather as a young man.

The section on Matt Zarrella depended on numerous sources, including the following: "Soldier Gets Proper Burial After 39 years," by David A. Markiweicz, *Atlanta Journal-Constitution,* July 3, 2005: 2D; "Dogs to Aid in Search for Vietnam MIAs," by Matt Sedensky, *Associated Press,* February 11, 2003, http://www.armytimes.com/legacy/new/1-292925-1582144.php, accessed July 2012; and "Is Time Running Out to Find Soldiers' Remains in Vietnam?" by Geoffrey Cain, *Time,* May 13, 2011, http://www.time.com/time/world/article/0,8599,2071021,00.html, accessed January 2012. Mary Healey Jamiel's documentary on Zarrella's work, *Reliance,* is forthcoming. You can access trailers, photographs, and more information on the website: http://www.reliancethemovie.com/index.html.

14: Running on Water

Interviews and correspondence for this chapter include those with Mike Baker; Roy and Suzie Ferguson; Kevin George; Lisa Higgins and her granddaughter Haylee; Kathy Holbert; Nancy Hook; Marcia Koenig; Roxye Marshall; Paul Martin; Joe Mayers of the Louisiana Search and Rescue Dog Team; Lisa Mayhew; and Andy Rebmann. I was also able to observe handlers and trainers at several seminars and sheriff K9 trainings.

Books and articles important to this chapter include *Cadaver Dog Handbook,* especially pages 151–162, on water recovery, as well as Marcia Koenig's piece, "Water Search on the Iowa River," *NASAR Newsletter* 5, no. 5, 1988: 11–15.

A podcast of Lisa Higgins talking to cadaver-dog handlers, "EPISODE19—NSDA-POD Cast Water Human Remains Recovery" (January 23, 2011), is available at http://recordings.talkshoe.com/TC-21763/TS-438866.mp3.

In addition to interviews, I used press accounts for the drowning cases in this chapter. Although many news accounts used victims' names, I chose not to identify them for privacy reasons and lack of relevance to this project. Material for the section on Afghanistan, along with interviews, included a PowerPoint presentation about the recovery that included photographs of Strega working the river; the written report on the recovery; and news articles from the Associated Press and the BBC, among others.

15: The Perfect Tool

Interviews and correspondence for this chapter include those with Joan Andreasen-Webb, Ken Furton, Paul Martin, Nick Montanarelli, and Arpad Vass.

This chapter depended greatly on news accounts, scientific articles, and technical reports, patent applications, and grants on animals and machines, including "The Use of Arthropods as Personnel Detectors," by Clyde S. Barnhart Sr., Defense Technical Information Center, U.S. Army Land Warfare Laboratory, Aberdeen Proving Ground, Maryland (August 1968); "Mimicking the Human Smell Sensing Mechanism with an Artificial Nose Platform," by Sang Hun Lee, Oh Seok Kwon, Hyun Seok Song, Seon Joo Park, Jong Hwan Sung, Jyongsik Jangb, and Tai Hyun Parka, *Biomaterials* 33, 2012: 1,722–1,729; "JIEDDO: The Manhattan Project That Bombed," by Peter Cary and Nancy Youssef, Center for Public Integrity Report (March 27, 2011; updated August 10, 2011), http://www.publicintegrity.org/2011/03/27/3799/jieddo-manhattan-project-bombed; "Technology Falls Short in the War Against IEDs," by Sandra Erwin, *National Defense Magazine* blog post (October 20, 2010), http://www.nationaldefensemagazine.org/blog/lists/posts/post.aspx?ID=221, accessed February 2012.

The section on the disastrous "search vultures" of Germany depended on reporting from *Der Spiegel*, ABC News, the United Kingdom's *Daily Mail*, and *Popular Science*, among other news media. The peer-reviewed research on avian olfaction is somewhat limited. UCLA professor emeritus of physiology Bernice Wenzel noted with pluck that avian olfaction "has been a small but persistent research topic for the last half century" in her piece "Avian Olfaction: Then and Now," *Journal of*

Ornithology 148, 2007: 191–194. David Malakoff provided an overview of the topic in *Science* magazine: "Following the Scent of Avian Olfaction," *Science* 286, no. 5440, October 22, 1999: 704–705, http://www.sciencemag.org/cgi/content/full/286/5440/704, accessed May 2012. Famous ornithologist Kenneth Stager reported the role of turkey vultures in the natural gas industry in the 1930s in his publication *The Role of Olfaction in Food Location by the Turkey Vulture* (Cathartes Aura) (Los Angeles: Los Angeles County Museum, 1964). The article from German scientists using rabbit corpses for their study was titled "An Electronic Body-tracking Dog?" by C. Hädrich, C. Ortmann, R. Reisch, G. Liebing, H. Ahlers, and G. Mall, *International Journal of Legal Medicine* 124, 2010: 43–47. The FBI study on clandestine graves is "Advanced Scientific Methods and Procedures in the Forensic Investigation of Clandestine Graves," by Daniel O. Larson, Arpad A. Vass, and Marc Wise, *Journal of Contemporary Criminal Justice* 27, no. 2, 2011: 149–182.

The section on Vass's sniffer machine, the LABRADOR, used the following: "LABRADOR: New Alpha Dog in Human Remains Detection?" by Douglas Page, *Forensic Magazine* (June 10, 2010), http://www.forensicmag.com/article/labrador-new-alpha-dog-human-remains-detection, accessed January 2012; and "A New Forensics Tool: Development of an Advanced Sensor for Detecting Clandestine Graves," by Arpad Vass, Cyril V. Thompson, and Marc Wise, U.S. Department of Justice Grant Final Report no. 231197, July 2010, www.ncjrs.gov/pdffiles1/nij/grants/231197.pdf, accessed January 2012.

16: Grave Work

Interviews, correspondence, personal communication, on-site observation, and personal experience for this chapter include those with Mary Cablk; Chris Chia; cadaver-dog handler Ann Christensen; Edward David; genealogist Pat Franklin; cadaver-dog handler Charm Gentry; Suzi Goodhope; cadaver-dog handler Gwen Hancock; Lisa Higgins; Kathy Holbert; cadaver-dog handler Lisa Lepsch; genealogist May MacCallum; Paul Martin; Thomasville assistant city manager Kha McDonald; Deborah Palman; Thomas County Historical Society curator Ephraim J. Rotter; Mississippi archaeologist John M. Sullivan; Arpad Vass; and Jefferson County, Kentucky, coroner Barbara Weakley-Jones.

Nehemiah Cleaveland, author of the epigraph, was a tutor and historian of Bowdoin College. His book, *Green-wood Illustrated* (New York: R. Martin, 1987), is on the rural cemeteries of America, http://books.google.com/books/about/Green

_wood_illustrated.html?id=Zd4TAAAAYAAJ, accessed June 2012. One of the most evocative and helpful academic articles for this chapter was Cornell University history professor Aaron Sachs's work "American Arcadia: Mount Auburn Cemetery and the Nineteenth-Century Landscape Tradition," *Environmental History* 15, no. 2, 2010: 206–235.

The Old Spanish Trail research is a work in progress. Mary Cablk and Barbara Holz-Montemayor have done work on one site there: "Preliminary Findings from Research at a Lander Site in Southern Nevada," Desert Research Institute report, June 8, 2010. Mary is continuing that research and working on permits to excavate.

A group of archaeologists and cadaver-dog handlers are also working in several areas where prehistoric mound builders created the North American equivalent of the pyramids. There's a great little video, "The Lost Worlds of Georgia," written and directed by Gary C. Daniels, that shows the Mississippian mounds in the Southeast, at http://www.youtube.com/watch?v=TuAlBcmKbPY. The following were also helpful: "Cadaver Dogs as Part of the Archaeological Survey Process, Preliminary Results from Phase I: Mississippi Spring 2011," by Paul Martin; "Best Practices for the Use of Cadaver Dogs to Locate Cold Case, Historical, and Pre-historical Burials," by Paul Martin and John M. Sullivan, paper presented at the National Association for Search and Rescue National Conference, Little Rock, Arkansas, 2009; and "Using Canines as a Remote Sensing Tool: What Archaeologists Can Learn From SAR Dogs," by Heather Roche, unpublished paper, May 2005.

The section on Thomasville prison depended on, among other sources, notes from Ephraim Rotter, which included a short account of the twelve-day prison camp, "Prisoner of War Camp/Histories Folder"; William Smith Brown, in his *Harper's New Monthly Magazine* article, "The Winter Climatic Resorts of Three Continents," January 1887: 868–876, extolled the virtues of Thomasville, Georgia. Lessel Long's moving account of Andersonville and Thomasville, *Twelve Months in Andersonville: On the March—In the Battle—In the Rebel Prison Pens, and at Last in God's Country* (Huntington, IN: Thad and Mark Butler, 1886), can be accessed through the Library of Congress: http://archive.org/details/twelvemonthsinan01long. A few photos of the current site of the Thomasville prison camp can be found at http://www.civilwaralbum.com/misc12/thomasville1.htm. J. David Hacker's quote comes from his Opinionator blog post in the *New York Times*, "Recounting the Dead," September 20, 2011, http://opinionator.blogs.nytimes.com/2011/09/20/recounting -the-dead/, accessed June 2012.

One of the most visible teams working on historic remains is the Institute for Canine Forensics, http://www.k9forensic.org/.

The section on West Virginia included Creighton Lee Calhoun's research on Southern apples, *Old Southern Apples, Revised and Expanded Edition: A Comprehensive History and Description of Varieties for Collectors, Growers, and Fruit Enthusiasts* (White Junction, VT: Chelsea Green Publishing, 2010). Historian Wilma A. Dunaway looked at slavery in the mountain states in her book, *The African-American Family in Slavery and Emancipation* (Cambridge University Press, 2003), 13.

The section on South Carolina depended on numerous online genealogies and ancestry records, as well as family letters and lore.

17: A Second Wind

This chapter depends greatly on personal observations and communications but also on a relatively new field of research best described as working-canine epidemiology. Interviews and personal communications include those with Mike Baker; Dr. Wendi Dick, director of environmental health, pre-9/11 and post-deployment, with the Office of Public Health, in the Veterans Health Administration; Dr. David F. Goldsmith, associate research professor of environmental and occupational health at George Washington University; retired Durham Police K9 Officer Danny Gooch; Deak Helton; Kathy Holbert; Nancy Hook; former K9 Officer Sean Kelly; Larry Myers; Dr. Michael Peterson, chief consultant post-deployment health, Office of Public Health, in the Veterans Health Administration; and Steve and Sandy Sprouse.

Among the articles, conference papers, and poster sessions helpful for this topic was "Animal Sentinels for Environmental and Public Health," by John S. Reif, *Public Health Reports* 126, supplement 1, 2011: 50–57; the foundational article on military working-dog health is "Excess of Seminomas Observed in Vietnam Service U.S. Military Working Dogs," by Howard M. Hayes, Robert E. Tarone, Harold W. Casey, and David L. Huxsoll, *Journal of the National Cancer Institute* 82, no. 12, 1990: 1,042–1,046; "Military Working Dogs Are Sentinels for Exposures and Disease Risks Among Deployed Veterans," by David F. Goldsmith, poster session for the International Society for Exposure Science, October 23–27, 2011, Baltimore, Maryland; "Incidence of Zoonotic Diseases in Military Working Dogs Serving in Operations Desert Shield and Desert Storm," by Kay D. Burkman,

George E. Moore, and Michael R. Peterson, *Military Medicine* 166, no. 2, 2001: 108–111; "Medical Surveillance of Search Dogs Deployed to the World Trade Center and Pentagon: 2001–2006," by Cynthia M. Otto, Amanda B. Downend, George E. Moore, Joanne K. Daggy, D. Lauren Ranivand, Jennifer A. Reetz, and Scott D. Fitzgerald, *Journal of Environmental Health* 73, no. 2, 2010: 12–21; "Pathology and Toxicology Findings for Search-and-Rescue Dogs Deployed to the September 11, 2001, Terrorist Attack Sites: Initial Five-Year Surveillance," by Scott D. Fitzgerald, Wilson K. Rumbeiha, W. Emmett Braselton, Amanda B. Downend, and Cynthia M. Otto, *Journal of Veterinary Diagnostic Investigation* 20, 2008: 477–484; *Polluted Pets Summary,* by Olga Naidenko, Rebecca Sutton, and Jane Houlihan, an Environmental Working Group report, April 2008, http://www.ewg.org/reports/pets, accessed August 2012; and "Do Environmental Pollutants Cause Cancer in Dogs?" by Sophia Yin, *The Bark* 44, September/October 2007, http://thebark.com/content/do-environmental-pollutants-cause-cancer-dogs, accessed June 2012.

18: Wag

This chapter includes interviews, personal communication, and correspondence with Joan Andreasen-Webb; Mike Baker; Kathy Holbert; Nancy Hook; retired Durham Sheriff K9 Sergeant Rick Keller; Lisa Mayhew; Steve Sprouse; and members of the Durham Sheriff and Durham Police Department K9 units.

A textbook that was enormously helpful for this chapter was *Dog Behaviour, Evolution, and Cognition*, by Ádám Miklósi (Oxford: Oxford University Press, 2007). Miklósi is head of the Department of Ethology at the Eötvös University in Budapest, and one of the pioneers of modern dog research.

The Patricia McConnell quote on sex differences appeared in her thoughtful piece "What Are the Differences Between Male and Female Dogs?" in *The Bark,* April 15, 2009, http://www.thebark.com/content/gender-gap, accessed August 2012.

The work that Temple Grandin did in her lab at Colorado State University on cattle and whorls on their foreheads, "Cattle with Hair Whorl Patterns Above the Eyes Are More Behaviorally Agitated During Restraint," by Temple Grandin, Mark J. Deesing, J. J. Struthers, and Ann M. Swinker, appeared in *Applied Animal Behaviour Science* 46, 1995: 117–123. Lisa Tomkins riffed off that work with "Lateralization in the Domestic Dog (*Canis Familiaris*): Relationships Between Structural, Motor, and Sensory Laterality," by Lisa M. Tomkins, Kent A. Williams, Peter C. Thomson, and Paul D. McGreevy, *Journal of Veterinary Behavior* 7, no. 2,

2012: 70–79; and "Behavioral and Physiological Predictors of Guide Dog Success," by Lisa Tomkins, Peter C. Thomson, and Paul D. McGreevy, *Journal of Veterinary Behavior* 6, no. 3, 2011: 178–187. Some video and explanation of Tomkins's work can be viewed on the archives of the *Catalyst* show of Australia's ABC: "Left Paw Right Paw," http://www.abc.net.au/catalyst/stories/3465535.htm.

Some early shots of Kathy Holbert's litter of pups, including Coda, can be found at http://www.chiodokennels.com/chiodo_kennels/REZA_ARKO_PUPPY _PAGE.html.

On imitative behavior among corvids, see, for example, "Social Learning Spreads Knowledge About Dangerous Humans Among American Crows," by Heather N. Cornell, John M. Marzluff, and Shannon Pecoraro, *Proceedings of the Royal Society B* 279, 2011: 499–508. Friederike Range of the Department of Cognitive Biology, University of Vienna, and head of the Clever Dog Lab at the Messerli Research Institute, University of Veterinary Medicine, Vienna, was among the first researchers to look at imitation or social learning in dogs; see, for example, "Automatic Imitation in Dogs," by Friederike Range, Ludwig Huber, and Cecilia Heyes, *Proceedings of the Royal Society B* 278, 2011: 211–217.

The other group studying social learning in dogs at Eötvös University in Budapest has authored such articles as "Dog as a Model for Studying Conspecific and Heterospecific Social Learning," by Enikö Kubinyi, Péter Pongrácz, and Ádám Miklósi, *Journal of Veterinary Behavior* 4, no. 1, 2009: 31–41; and "When Rank Counts—Dominant Dogs Learn Better from a Human Demonstrator in a Two-Action Test," by Péter Pongrácz, Petra Bánhegyi, and Ádám Miklósi, *Behaviour* 149, 2012: 111–132.

The South African study that I found so evocative—"Early Prediction of Adult Police Dog Efficiency—A Longitudinal Study," by J. M. Slabbert and Johannes S. J. Odendaal, *Applied Animal Behaviour Science* 64, 1999: 269–288—has not been replicated, to my knowledge.

Index

INDEX

INDEX

INDEX

Renzo (German shepherd), 240
retirement: of dogs, 250–52
Revolutionary War, 233, 243–48
rewards
    accidental, 94–95
    Baker's training program and, 117
    for bees, 48
    for bloodhounds, 41–42
    false alerts and, 147, 148
    food as, 166, 219
    "go out" and, 149
    and Hook's training of Solo, 70, 71, 72
    importance of, 115
    for lack of fear, 96
    negative searches and, 152, 153
    and personality of dogs, 186
    self-, 253
    for Solo, 70, 71, 72, 101, 116, 134, 135, 136, 142, 143, 147, 148, 152, 153, 209, 247, 248, 253
    toys as, 101, 116, 117, 134, 135, 136, 142, 143, 152, 153, 165–66, 209, 248
    for "Wasp Hound," 219
    weaning dogs off, 165
    for working dogs, 115
Reza (German shepherd), 275
Rhodes, Robin, 187–88
Rick (German shepherd), 123–24, 125
Rin (German shepherd), 250
Robby's Law, 259
Rocco (shepherd), 240
Rotter, Ephraim, 236
Rufus (German shepherd), 61, 62, 186

sagdid (Zoroastrian rite), 20–21
Sagebiel, John, 89
"sanitary dogs," 190
Sanson, Greg, 199, 214
Sassoon, Siegfried, 189, 190
Scamp (schnauzer), 23
scavengers, 20, 24
scenarios, 161–67, 170–74
scent
    of animals, 61
    artificial nose and, 220–21
    of buried bodies, 87
    comparison of human with dog, 169
    of death/decomposition, 18, 21, 61, 80–82, 83, 84–89, 96, 200
    dispersal of, 138–39
    of historic human remains, 236, 240, 242
    human body, 60–61
    human remains versus live human, 138
    ingested chemicals and, 87
    longevity of, 185
    and role of scent evidence in court system, 154–55
    submerged bodies and, 202
    training and, 60–61
    "The Wasp Hound" patent and, 219–20
    water searches and, 209, 210, 212
    See also working scent

scent drive: of Solo, 10
scent lineup, Pikett's, 155–56, 157
Schatzie (German shepherd), 166–67, 205
Schutzhund (ring sport), 3, 123–24
Search & Rescue Dog training packages, 93
search patterns, 103–4, 116
search-and-rescue dogs, 17–18, 34, 44, 54, 115, 195, 257
searches
    and animals in search areas, 183
    blind, 152–53
    Cat's reactions to, 176–80
    comparison of land and water, 209–10, 211
    in difficult terrain, 182
    dogs' unwillingness to stay within confines of, 166–67
    handlers' preparation for, 181–83
    learning from, 177–78
    negative, 152–53, 177
    options for outcomes of, 176–77
    reality of, 178–79
    time frame for, 184
    types of areas for, 183–85
    unreasonable, 151
    See also specific search or type of search
second-dog syndrome, 268
Seminole Indian Wars, 191
sentinel animals, 257–58
Sepot, Peter, 119, 128–29
September 11, 2001, 193, 256
Shakespeare, William, 39, 161, 164
sheep: scent of dead, 86
shell game, 75–79
Shepard, Paul, 15, 25
shepherds, 52–53, 59, 267. See also type of shepherd or specific dog
Sherlock (vulture), 227
Shiraz (Belgian Malinois), 107, 236
Shubin, Lester, 37–38, 49
Sinai War (1972), 192
singleton dog(s), 3, 5–6, 13, 102
skunks, 51
slave searches, 43–44, 235, 237–41, 245, 246
Sledge, Michael, 191
Slick, Tom Jr., 48, 65
Smalley, Barb, 7–8, 279
smell, sense of
    and aging of dogs, 254
    bears and, 30–31
    bloodhounds and, 30, 37, 38–39, 40
    brain and, 84, 87
    of Cat, 84–85
    chocolate studies about, 32
    cognition and, 35–37
    comparative studies about, 28, 29–31, 36–37
    of hunting dogs, 33
    identical twin study and, 35–36
    importance of, 29
    lack of knowledge about dogs, 28
    medical importance of, 31
    number of receptors and, 37

About the Author

Cat Warren, an associate professor at North Carolina State University, has taught science journalism, editing, media theory, and gender studies there since 1995.

In her first career as a newspaper reporter, she covered crime, poverty, the environment, and politics at newspapers across the United States. She won numerous state and regional press association awards for that work, starting in 1985 through 2004. Her reporting ran nationally on numerous occasions, on subjects as diverse as former Secretary of the Interior James Watts's efforts to profit from Native American oil reserves in Wyoming to homophobic teenagers preying on gay men in Connecticut. She was on the special projects desk at the *Hartford Courant* when she decided to take a break from newspapers to substitute-teach an editing class at a university in Oregon. She fell in love with teaching. Instead of returning to newspapers, she went back to school in her mid-thirties to get her doctorate.

Cat has kept her hand in magazine and newspaper writing and editing. She worked at University of Illinois Press while finishing her doctorate, and was the editor of a national higher-education magazine, *Academe*, for three years. She has published academic articles and edited two academic books. She has also taught at the University of Portland, Stanford University, and Duke University.

Cat was born in Oregon, where her father taught her to fly-fish, to collect snakes in the summer and release them in the fall to hibernate, and to identify the trees and plants of the Pacific Northwest. Once she got a couple decades of rootlessness out of her system, she settled in North Carolina. She lives in Durham with her husband, David Auerbach, a baker of bread and a professor of philosophy at North Carolina State University; their two German shepherds, Solo and Coda; and two prolific beehives.

What the Dog Knows is her first book. For more information, visit her website at catwarren.com, where you'll find links to follow her on Twitter and Facebook.